Grow anything a

The GARDEN DOCTOR

Book 3

Jacob R. Mittleider, Litt.D.

Originator of "growbox" gardening

Author of:
MORE FOOD FROM YOUR GARDEN
FOOD FOR EVERYONE
GROW-BED GARDENING
LET'S GROW TOMATOES
GARDENING BY THE FOOT

AUTUMN HOUSE® PUBLISHING COMPANY
P.O. Box 1139, Hagerstown, Maryland 21741-1139

Published and distributed by
Autumn House® Publishing Company
Hagerstown, Maryland

Copyright © 1990 by Jacob R. Mittleider
All Rights Reserved

This book or any part thereof may not be reproduced in any manner whatsoever, except as provided by law or as brief excerpts in critical reviews or articles, without the written permission of the author.

Cover photo: Superstock
Cover photo inset: Comstock/R. Michael Stuckey

Autumn House Cataloging
Mittleider, Jacob R.
  The garden doctor.

  3 vols.
  1. Plants—Diseases. 2. Fruit—Diseases and pests.
3. Vegetables—Diseases and pests. I. Title.

     631.8

ISBN 1-878951-02-5
3-vol. set ISBN 1-878951-03-3

Printed in the United States of America

Dedication

While preparing the manuscripts for the three volumes entitled *The Garden Doctor,* I was often made aware of the influence that Dr. J. R. Wise and Douglas Dietrich, our two sons-in-law, had on the circumstances that made the research and hands-on experience possible.

Their stalwart and noble characters, their honesty, compassion, hard work, perseverance, and morality, had a direct influence on the scope and quality of work we have pursued.

Not once did family circumstances give reason for concern because of domestic or financial problems. Their children have high goals and are pleasant and respectful.

These three volumes have been prepared with a prayer to be a help in food production for people around the world. Wherever a degree of success is achieved, Jim and Doug share a large part.

THE AUTHOR

A Word to the Reader

For you to gain the most benefit from *The Garden Doctor,* it is important that you understand what these books are intended to do—and what they are not intended to do.

Obviously, you are interested in gardening. Perhaps you are a newcomer, a novice, and want to know where and how to begin. Or perhaps you have had some experience and you want to improve your skills, increase production, make your work easier. Then again, perhaps you want to know how to apply techniques that work well in a small kitchen garden on a larger scale, perhaps a truck farm.

The author, Dr. Jacob R. Mittleider, has done all these, and he can help you. Following 20 years of experience growing flowers and vegetables commercially, he embarked on a program of sharing his expertise with gardeners and would-be gardeners around the world. His methods have changed agricultural history in the United States, South Pacific, Japan, and Africa. At presstime he is under appointment to the U.S.S.R. to lecture and demonstrate.

Fortunately, Dr. Mittleider doesn't just teach. He also writes, so that many more thousands of people can benefit from his experience. One of his first books was *Food for Everyone,* written with the help of his friend Andrew N. Nelson, Ph.D. It covered all aspects of growing healthy and nutritious, disease- and pest-resistant garden crops. It included detailed instructions and color photographs of how to prepare the soil, select the seed, even how to store or market the produce. This book related how Dr. Mittleider's methods have proved successful even in areas formerly thought unusable.

Other books followed. *Mittleider Grow-box Gardens* detailed how to build frames to hold "homemade soil." These frames, Dr. Mittleider's specialty, helped to provide good plant nutrition, aid in the control of weeds, and make cultivation easier. *Grow-bed Gardening* applied the same principles to larger growing areas where boxes were not practical. Then there was *Let's Grow Tomatoes*—142 pages of practical information on growing this vegetable that is such a favorite around the world.

If you are a novice gardener, or want to try the Mittleider method, we highly recommend these books, any one or all. To find out how to obtain them, write to this publisher.

The Garden Doctor is a different type of book. It does not purport to cover the entire range of gardening, but focuses on the soil, the basic foundation for good plants. What makes good soil? How can you tell if the soil is poor, and how do you improve it? How do various symptoms in the plant reflect deficiencies in the soil makeup?

Dr. Mittleider discusses elements, minerals, and soil components. He tells why they are important, what to do when they are lacking, and when to stop worrying about them.

No matter how much experience you have had in gardening or farming, you can profit from these books. Not only will they help you increase productivity; they will also help you get more pleasure out of gardening. You will find out how much fun it is to get your hands dirty!

Contents

BORON (B)
A Micronutrient — 15

Section 1:	Definition of Boron	15
Section 2:	Chemistry of Boron in Plants	16
Section 3:	Chemistry of Boron in the Soil	19
Section 4:	Forms of Boron Utilized by Plants	22
Section 5:	Actions Affecting Boron Supply	26
Section 6:	Methods of Applying Boron	28
Section 7:	Symptoms of Boron Deficiency	30
Section 8:	Symptoms of Boron Excess	32
Section 9:	Cautions Regarding the Use of Boron	33
Section 10:	Special Suggestions Regarding Boron	35
Section 11:	Summary and Review of Boron	37

MOLYBDENUM (Mo)
A Micronutrient — 49

Section 1:	Definition of Molybdenum	49
Section 2:	Chemistry of Molybdenum in Plants	50
Section 3:	Chemistry of Molybdenum in the Soil	53
Section 4:	Forms of Molybdenum Utilized by Plants	55
Section 5:	Actions Affecting the Supply of Molybdenum	56
Section 6:	Methods of Applying Molybdenum to Plants	58
Section 7:	Symptoms of Molybdenum Deficiency	60
Section 8:	Symptoms of Molybdenum Excess	61
Section 9:	Cautions Regarding the Use of Molybdenum	62
Section 10:	Suggestions Regarding Molybdenum	63
Section 11:	Summary and Review of Molybdenum	64

Salinity (Soil pH) Symptoms of Distress — 69

Two or More Deficiencies — 75

Miscellaneous Symptoms — 101

Garden Grow-Boxes — 120

How to Make Grow-Boxes — 124

Using Custom-made Soil — 131

Index — 135

Foreword

Recognizing deficiency symptoms is a noble achievement, but of equal importance is knowing the right amount of a nutrient or compound to satisfy the needs of growing plants.

Animal lovers take pleasure in the appearance and performance of their animals. They provide clean, sanitary living quarters and a balanced diet. Even so, abnormal symptoms of deficiencies can and do develop. After diagnosing the symptoms and determining that a mineral deficiency exists, they supply supplements of the limiting factors promptly. To a large degree, animals are what they eat.

People enjoy life best when they are healthy. They too are what they eat. But even those enjoying the best and most nutritious eating habits frequently must supplement the mineral and vitamin supply in their diet in order to maintain radiant health.

Animals and people are living creatures. After many years of research and experimentation, considerable information is available on the functions various minerals perform in man and animals, and also how a deficiency adversely affects bodily functions.

Plant food crops are living things! They too are what they eat. The same environmental conditions that cause deficiency symptoms in man and animals exist for plants. The nutrient requirements are the same.

Radiant health in animals and people is obvious just by observation. A healthy family garden or field crop is the dream of every grower.

Unfortunately, it is no longer possible just to prepare a good seedbed, plant good seed, water, cultivate, and reap a uniform overall bountiful crop. Nutritional soil deficiencies are here to stay. In fact, they are increasing every year. But the notion that soils are worn out has no basis! It is simply impossible to wear out soils!

Rocks are the parent materials of soils. They are the same thing! Soils are composed of metals (chemicals to the chemist; minerals to the naturalist), elements, compounds, etc. Water-soluble materials tend to move with the soil solutions.

There are two natural elements that affect the metal chemical supply of soils—wind and rain.

In one sense soils are alive. Chemical changes are continuous—from insoluble compounds to soluble and vice versa. Soluble materials move with the soil solution both in and out of the soil. Soil moisture is the source of drains, creeks, lakes, and rivers. These eventually reach the oceans.

Plants extract their food from the soil solution. Soils that are well fortified with the essential plant chemicals (metals, elements, compounds) produce healthy quality crops. Centuries of rain and snow have leached the land area of the world and reduced (diluted) the chemical supply. Considerable quantities of the chemicals that originally were part of the soil structure are now deposited on the floors of oceans around the world. Several countries are mining the oceans to extract the common and rare metals.

As the chemical supply of the soil is reduced through losses from leaching and crop removal, quality and yields decline.

Rocks containing insoluble natural chemicals abound throughout the world. Man has learned how to use the products of the rocks to increase the soil supply of the essential plant foods.

The key to healthy quality crops is fertile soil!

The three books entitled *The Garden Doctor* have been prepared to help you understand, through the visual appearance of plants and crops, whether distress symptoms do appear, and if so, the corrective treatment to use in eliminating the problem.

What They Say...

National Courier

"Dr. Jacob Mittleider knows one thing that can change the face of the earth, and he's trying to get someone to listen. No matter how poor man or man's soil is, if there's a way to get water to it, with only a plot of ground 20′ x 100′, a man can feed his family better than it has ever been fed before.... The Idaho baker-turned-agricultural wizard specializes in turning what is known as 'devil land' into unbelievably productive land."

San Juan *Record*

"Dr. Jacob Mittleider . . . has shown that food can be produced 'in the world's worst soil.'"

The Chronicle
Bulawayo, Zimbabwe, Africa

"Dr. Mittleider is performing what many people would have thought to be humanly impossible—developing a vegetable garden on the infertile sandy soil around Solusi College."

Trinidad Guardian

"Desolate desert areas have been converted to thriving areas of vegetation, and rocky barren areas have been brought under successful food production, thanks to a revolutionary new method devised by a United States agriculturist.

"At a time when starvation and declining food production have become issues of urgent international importance, Dr. Jacob Mittleider has performed a feat of 'agricultural wizardry.'"

Izvestyia

"Everywhere, growing crops on the most diverse and sometimes useless soils and even in the boxes filled with sand and sawdust mixtures, Dr. Mittleider created his miracle. Weeds vanished, crops grew bigger; the highest quality yields turned several times bigger."

Utah Navajo-Baa Hane

"Dr. Mittleider has shown in his demonstration garden project at Halchita that it is possible to grow a successful garden despite poor soil conditions. . . . The garden was started in June. In four weeks the barren, rocky plot was transformed into a lush garden that has produced a variety of crops."

The Social Industry

"Professor Jacob Mittleider has created the gardening technology for both greenhouses and outdoors that makes it possible to get yields from three to ten times higher than when traditional methods are used. This method has been tested throughout the world—from Upper Volta to Japan. It has been proved that neither rocky soil nor high temperatures nor cold can prevent one from getting high yields."

Charles A. Jones
On behalf of the Church of Jesus Christ of Latter-day Saints

"My personal feelings and those shared by my associates are that the Mittleider Method is a most enjoyable means of growing food. . . . The results are fantastically more rewarding than I have ever had in my 30 years of growing vegetables by conventional means. The production is heavier, quality higher, and the results more dramatic in every way. I am pleased to recommend the Mittleider Method to anyone either for pleasure or profit."

Sonny Ramdath
Farmer, Trinidad and Tobago, West Indies

"His method is simple, based on the laws of nature, yet very invaluable, and can be applied any time, any place, under any conditions—with sure results. With my experience in the field of agriculture, I can safely say that this man knows his stuff, and he sure is an asset to the world."

Preface

Plant symptoms originating from nutrient deficiencies are easily verified and/or obtained through clear culture solutions and inert growing media in laboratory experiments.

In such experiments, nutrients established thus far to be essential for the best performance of plants are incorporated in the growing medium, with one exception.

One nutrient of the 13 is left out of every trial experiment. Through this method accurate symptoms for each nutrient can be produced and documented. Such experiments are generally conducted in an artificial, closed environment. The results can be very accurate.

Normal field conditions, however, are quite different. Farm land has many variables, such as virgin supplies of essential nutrients, variables in pH, temperature changes, moisture levels, and aeration. Nevertheless, a plant's response is determined largely by the environment.

A crop or a plant is the most valuable and accurate "patient" a grower has. Its appearance, growth, and fruit reveal whether problems exist or if all is well.

Every one of the 13 essential nutrients that man can regulate affects the crop in a slightly different way. There are also miscellaneous factors that affect plant performance.

The materials on nutrient deficiencies, including salinity and miscellaneous factors that affect plant performance, have been prepared to help you recognize, analyze, and prescribe appropriate treatment to produce high-quality and high-yield crops.

All deficiencies were documented under field conditions. Corrections were made and the changes documented as they occurred.

In comparing field-condition symptoms with controlled laboratory experiments, there will be differences. Therefore, plant response to corrective treatments have been included to verify diagnosis.

The Author

The predominant symptom of boron deficiency is death of the terminal buds.

The crinkled, riddled cotyledon bean leaves and missing terminal bud indicate boron deficiency.

BORON
B

A Micronutrient Deficiencies and Script

Section 1:
DEFINITION OF BORON

The chemical symbol is a capital B.

Boron is a nonmetallic element present in borax and is obtained either as an amorphous (lacking definite form; shapeless) or crystalline form reduced from its compounds.

The boron used in the manufacture of freezers, stoves, refrigerators, kitchen sinks, and glassware in a modern kitchen is enough to produce 16 tons of alfalfa hay.

For nearly 400 years boron has been used as a fertilizer. But only since 1915 has boron been recognized to be an essential nutrient for plant growth.

Seawater contains between 4.5 and 5.0 ppm of boron.

Boron—B

Boron is an essential nutrient. The twisted, deformed bean leaves indicate boron deficiency.

The dead leaf portions on the bean plants suggest boron deficiency.

When bean pods are crooked and develop a scar seam on the inside of the curved area, it is boron deficiency.

Section 2:
CHEMISTRY OF BORON IN PLANTS

Boron performs at least 15 functions in plant growth.

It affects the flowering and fruiting processes, pollen germination, cell division, the metabolism of nitrogen and carbohydrates, active salt absorption, hormone movement and action, the metabolism of pectic substances, water metabolism, and the functions of water in plants.

In addition, it is believed boron is a constituent of membranes to function in precipitating excess cations, which act as buffers.

Boron functions in the maintenance of the conducting tissues and exerts a regulatory effect on other elements within the plant. Also, there is strong evidence that boron is necessary for translocating (moving) sugar within plants.

In order for plants to function normally, they must absorb boron throughout their life cycle. This supports the position that new boron is required as new cells are formed.

It is apparent that boron moves very little inside a plant.

The amount of boron required by plants at any one time is very small.

The deficiency symptoms of boron and calcium are very similar. This is strong evidence that the two elements are very likely related in plant functions and growth.

Whenever the proportion of calcium to boron becomes unbalanced in the plant because of a deficiency in boron, the growing tips (terminal buds) fail to develop properly.

Boron—B

The scorched leaf tips and wilting bean leaf indicate boron deficiency.

The chlorotic and riddled cotyledon leaves have boron deficiency.
Two weeks after correcting for boron the new leaves were normal and green.

After correcting for boron, a new terminal bud grew and the leaves were normal.

Even though there must be a balance between the supplies of calcium and boron, still the margin between boron deficiency and boron excess is very narrow.

Excess boron is toxic and frequently fatal to plants.

Boron increases crop yields and improves the quality of forage grasses, fruits, and vegetables.

The amount of boron required is small. Every ton of alfalfa hay contains about 1 ounce. One hundred bushels of corn contain about .4 ounces. Every ton of sugar beets contains about 2.5 ounces.

The assimilation of boron by the plant is associated to the concentration of other ions in the nutrient quotient of the soil solution.

When the calcium supply is low, plants have a lower tolerance for boron, and the reverse is also true. Thus, when the supply of calcium is high, there will be a demand for extra boron.

Potassium is another element that must be in balance with boron in plant growth.

Nitrogen-starved plants need less boron than plants that are well supplied with nitrogen. This may be because of the effect nitrogen has on plant growth.

Phosphorus-starved plants require more boron than plants that are well supplied.

The fact that plants need boron is well established. But the amount required is small—between 10 and 35 ppm (between 5 and 20 pounds per acre).

If boron is available, many species of plants will absorb much larger amounts than needed—even to the point of toxicity.

Notice the healthy green leaves have no deficiency.

Notice the healthy green bean leaves three weeks after correcting for boron.

Boron—B

Plants generally have a wide tolerance for many of the essential nutrients, but this is not so with boron. The margin is very narrow, and any excess is toxic.

The predominant symptom of boron deficiency is death of the terminal buds.

Boron is important for proper cell division in plant growth.

Excess boron (above .05 percent) is toxic to plants.

Some crops, such as beans and strawberries, are very sensitive to boron excess.

Crops such as alfalfa and tomatoes are tolerant to higher levels of boron.

Sometimes boron excess is experienced in arid (dry) regions.

Borax and sodium borate are the same material—compounds of boron.

As a general rule, 20 pounds of borax (sodium borate) per acre is adequate to meet the minimum boron requirements of many crops.

Boron deficiencies are easily corrected with boron salts.

Three weeks after applying boron all scorched leaf portions had disappeared.

Boron—B

Section 3:
CHEMISTRY OF BORON IN THE SOIL

Thousands of acres of agricultural soils contain only enough available boron to produce one or two tons of alfalfa hay.

The amount of boron in 2 tons of hay is enough to make a cookstove, an icebox, and a set of china.

Some crops are affected more from boron deficiency than are others. Crops that are commonly affected are alfalfa, clover, and sugar beets.

Following is a list of other crops that are frequently affected with boron deficiency: celery, beets, cauliflower, apples, grapes, pears, walnuts, filberts, sunflowers, asters, and other ornamental crops.

Boron deficiency is the cause for a disease called heart rot and dry rot in sugar beets and red beets.

Native supplies of boron vary between 20 and 200 pounds per acre in the first 6- to 8-inch layer of soil. Of this supply, a large proportion is in the form of the highly insoluble complex mineral called tourmaline.

It is important to know that the native supply of boron is not reliable as to the proportion of available boron for crop use. And less than 5 percent of the total inventory may be available for use by plants.

The boron problem is larger than just the manufacture and use of fertilizers.

For Example:

Irrigation waters in arid (dry) regions must be developed with care because boron is often present, both in arid soils and in the

This block of beans has no deficiency.

Three weeks earlier the beans had serious boron deficiency. Here they are normal.

And a rewarding crop was harvested over several weeks from the healthy plants.

Boron—B

irrigation waters used on them. This boron is usually water-soluble sodium borate, also called borax.

There are irrigation waters that carry so much boron that it cannot be used on crops.

When evaluating irrigation water for crop production, sensitive crops such as beans and nut and fruit trees are usually injured if the boron content exceeds .05 ppm.

A damaging amount for semitolerant crops such as sweet potatoes, cabbage, tomatoes, onions, etc., is 1.33 ppm. And tolerant crops such as sugar beets, alfalfa, etc., are injured if the amount is higher than .2 ppm.

Humid regions are generally low in boron because the water-soluble forms have been carried off in the drainage water over many years. The boron that remains is in the form of highly insoluble minerals.

The native supplies of boron in humid regions today is normally 25 to 100 ppm (about 50 to 200 pounds) per acre. And only about 1 percent of this supply is soluble—even in hot water.

Boron occurs in two compounds—organic and inorganic.

The inorganic forms—which are mainly calcium, magnesium, and sodium borates—result from the slow weathering of minerals containing boron. The microorganisms in the soil and plants utilize this boron in their life processes and change the boron compounds into organic forms.

The concentration of boron is higher in the topsoil than in the subsoil. Realizing this may help to explain why boron deficiency is greatest in periods of dry weather. It seems apparent that during periods of drought,

Boron — B

Because boron is water-soluble, it moves with the soil solution.

The dark-red and reddish color and poor leaf growth are the result of boron deficiency.

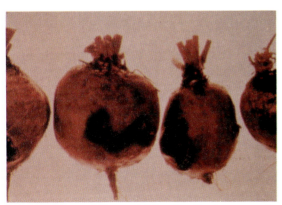

Beets can fail completely when the boron supply is lacking. The black lesions on the beet tubers is boron deficiency.

plant roots are forced to exploit only the lower soil horizons where the boron supply is very small.

Soils formed from igneous (rocks of volcanic origin) rocks are generally lower in boron content than soils derived from marine sediments.

Boron is perhaps man's oldest-known weed killer.

Boron can be leached from the soil.

Boron excess seldom, if ever, occurs in high rainfall areas.

Because there is a fine line between boron deficiency and boron excess, small amounts of boron salts should be applied until the requirements of the crop have been established.

Boron—B

Section 4:
FORMS OF BORON UTILIZED BY PLANTS

The availability of boron occurs in two broad forms—organic and inorganic.

Soil microorganisms and plants utilize inorganic boron and transform it into organic forms. When the microorganisms and plants complete their life cycles and die, the organic boron is oxidized back into inorganic boron.

Through the continuing weathering processes of unavailable forms and the boron made available for reuse by the microorganisms, the supply of boron in some soils remains adequate for crop growth.

The factors that result in insufficient supplies of boron on many soils are leaching losses, reversion to unavailable forms, and the higher requirements for boron of better crop varieties and improved cultural practices.

Deficiencies in boron are so common today that in the more intensively farmed areas of humid regions it is nearly standard practice to add 5 pounds of borax to each ton of mixed fertilizer.

Alfalfa has a high requirement for available boron, and as much as 100 pounds of borax may be added to each ton of fertilizer applied.

Broadcasting 10 pounds of boron per acre is usually safe for many crops. But if this amount is banded (placed in a narrow band) along the side of the planted rows, the plants may be injured.

In sandy soils especially, real precaution is necessary whenever borax is applied. The safer material to use is the less soluble mineral called colemanite.

Smooth, round bulbs with only one main taproot and red-tinged green leaves have no deficiency.

The quality of the beet leaves and the flavor are excellent on healthy plants.

After correction for boron, notice the healthy leaves. They have no deficiency.

Boron—B

Boron fertilizers in America come from mineral deposits in Death Valley, the Mojave Desert, and from the brine of Searle's Lake, all in California.

Sodium borate called borax and other sodium borates have been the most common materials used in agriculture. The percent of water-soluble boron in these borate compounds is 10.5 to 13.6 percent.

It should not be thought these are the only boron products available. Other boron minerals for fertilizers are being manufactured and sold.

For Example:

Colemanite calcium borate; this is less soluble than sodium borate.

Colemanite is recommended for use on sandy soils in high rainfall areas. It does not leach out of the soil as fast as sodium borate. Colemanite contains 10.1 percent of soluble boron.

Highly concentrated and highly soluble sodium borates have been developed for use in sprays and dusts. These are applied directly to the foliage of fruit trees, vegetable crops, and other crops in highly alkaline soils.

All boron materials can be mixed with regular fertilizers and applied together in one operation.

When deficiencies occur, the amount of available boron that can be used with safety depends on the crop, the soil, the season, the method of application, and the source of boron.

After correction of the soil supply for boron, the crop was normal.

Please notice the single taproots on the large smooth tubers.

Hollow stems on broccoli are from boron deficiency.

Boron—B

For Example:

Larger amounts of boron fertilizer can be applied to soils that have a high organic matter content to a high exchange capacity (or a high pH) than can be applied to light-arid soils of low organic matter content.

The application of borax fertilizers on alfalfa varies from 15 to 60 pounds per acre. The lower rates are generally applied on the light acid soils, and the heavier rates are generally used on heavier clay soils.

The position is quite firmly established that boron fixation is a rapidly reversible chemical process caused by alkalinity.

Several kinds of boron-containing fertilizers are available in different parts of the world, including:

1. Boronated gypsum.
2. Boronated calcium carbonate.
3. Boronated superphosphate.
4. Boronated calcium nitrate.

Boric acid can also be used either for application to the soil or applied in foliar sprays.

Boric acid is compatible with ammonium compounds.

Solubar contains 20.5 percent boron or 66.2 percent boron trioxide (an oxide containing three oxygen atoms). This product is a special formulation that is applied in solution.

Boron—B

A point to remember is that solubar is *not* compatible with oils.

Borax contains 11.3 percent solubar boron, or 36.5 percent boron trioxide.

More than 3 million tons of boron are used annually in agriculture. Nearly all is in the form of sodium borates.

Boron deficiency produces black lesions and cracks inside carrots.

These carrots have no deficiency.

A previous crop of carrots failed on the land. Boron was applied, and the crop has no deficiency.

Healthy carrots and assorted crops can be grown successfully on artificial soil (in this case, sawdust and sand).

The carrots were harvested from artificial soil. The bulbs are stubby with curved ends because the plants were transplanted by hand as very small seedlings. Transplanting affected the taproots.

Boron—B

Section 5:
ACTIONS AFFECTING BORON SUPPLY

Boron losses through crop removal are unavoidable. Nevertheless, such losses must be recognized and the boron replenished.

Every ton of alfalfa hay contains 1 ounce of boron. Each ton of sugar beets contains 2.5 ounces. One hundred bushels of peaches contains 4 ounces.

One hundred bushels of corn contain only .4 ounces of boron.

It is clear that the amount of boron removed from the soil by crop removal depends on the kind of crop and whether the crop is sold or fed on the farm and returned to the land as manure.

Boron losses from acid soils in humid regions are serious.

To a large degree, the texture of the soil controls the rate of boron movement in the soil.

Boron moves to lower depths in the soil profile in light-textured soils more than it does in heavier soils.

Among the many factors that influence the availability of boron is the pH of the soil.

Applying lime to the soil frequently results in lowering the availability of the soil boron.

The results gathered from research projects indicate that clay soils fix the most boron, silt loams fixed an intermediate amount of boron, and sands fixed very little boron.

The method recommended to convert fixed boron in alkaline soils to available boron is simply to lower the soil pH to mildly acid. This indicates that a reversible chemical reaction is involved whenever there is a

Boron—B

These medium-sized broccoli plants are normal and have no deficiency.

This broccoli plant has no deficiency. A quality head is developing nicely.

change in the soil pH from acid to alkaline and from alkaline to acid.

Another factor that influences the fixation of boron is the organic matter content of the soil. And still another factor is dry weather.

Dry weather hastens the appearance of boron deficiency symptoms. Therefore, extra precautions should be taken during the hottest and driest weeks of the growing season to ensure that boron supplies are adequate.

Boron deficiencies frequently show up on dairy farms first. This can be explained easily because the predominant crop grown is alfalfa and this crop requires high amounts of boron.

Vegetable crops deplete supplies of available boron quickly.

Irrigation water can intensify or improve boron deficiency problems. This depends on the amount of boron and the composition of the water.

Fertilizer management practices too can correct or intensify boron deficiency.

For Example:

If animal manures are returned to the land regularly, boron deficiency should seldom occur.

Using chemical fertilizers that contain traces of boron and applying these on a regular basis usually satisfies boron requirements.

Applications of boron on the soil give longer protection against deficiencies occurring than do foliar sprays.

Boron compounds are usually mixed with other fertilizer compounds before they are applied to the field.

Boron—B

Section 6:
METHODS OF APPLYING BORON

Boron products have been manufactured especially for use as sprays and dusts. These materials dissolve quickly and are highly concentrated. They are sprayed or dusted on the foliage of fruit trees, vegetable crops, and some field crops in alkaline soils.

Whether boron is applied to the soil or on the foliage as a dust or foliar spray, the results vary considerably.

The results obtained from using boron, whether satisfactory or bad, depend on the species of the plant; the soil-cultural practices, the rainfall, liming, and other factors including soil pH.

The usual mixture for applying boron as a foliar spray is 1 to 2 pounds boric acid mixed in 100 gallons of water.

For field application, borax (sodium borate) is commonly used. The borax is usually mixed with other fertilizer materials such as nitrogen, superphosphate, and potassium and broadcast on the surface of the field before it is plowed, or it is banded beside the rows at the time the seed is planted.

When top-dressing or side-dressing is practiced, borax can be mixed with the fertilizers used and applied during the various stages of plant growth.

These medium-sized cauliflower plants are normal and healthy.

This planting of cauliflower has no deficiency even though it is growing on hard clay soil, high in salinity.

Boron—B

In special cases, high-grade borate fertilizer compounds containing 14.6 percent soluble boron or 46 percent boron trioxide are recommended.

Please note, however, that this product should *not* be used as a source of boron for spray formulations, because some of the ingredients are *not* water-soluble. Neither should this material be used with ammonium salts, because of possible unpleasant chemical reactions.

Hollow curds of cauliflower are the effects from low boron supply.

If the deficiency is serious, the hollow areas develop a black, foul-smelling slime.

Also, the curds are often poorly filled; they are small and unmarketable.

Another boron deficiency symptom in cauliflower is rusty-colored, loose curds with leaves growing through the curd; flavor is bitter.

Mr. Buie is posing with his prize cauliflower curd.

Boron—B

Section 7:
SYMPTOMS OF BORON DEFICIENCY

Boron deficiency symptoms vary slightly with the type and age of the plant, the conditions under which it grows, and the severity of the deficiency.

The usual deficiency symptom that appears first is death of the terminal buds.

To the experienced grower, the first symptoms of deficiency may be only the collapse (sudden wilting) of a portion of a healthy mature leaf. The wilted portion does not recover—it dries up. The remainder of the leaf frequently is not affected.

As boron deficiency progresses, the symptoms are described as rosette- crown appearance. The rosette being multibranched terminal buds on a single growing tip.

Boron deficiency symptoms on young plants appear on the cotyledon (seed) leaves. These enlarge, thicken, have a leathery appearance and feel, a green-gray color, and are very brittle.

The leaves on boron-deficient plants tend to curl.

Petioles (the slender stock by which a leaf is attached to a stem) and sometimes the leaves become brittle.

Flowers may fail to form, or if formed, fruit may fail to set. The roots are generally stunted and stubby.

Boron deficiency in different species of plants can produce nonparasitic diseases.

Indeed, he had good reason to be proud of his production.

Boron—B

For Example:

Heart rot of beets; core disease of apples; brown rot of cauliflower; black stem of broccoli; and roan coloring of rutabaga.

Boron Deficiency Symptoms Summarized:

1. The early symptom is death of the terminal buds on the main stem. Later buds on side shoots die. Branches or shoots take on a rosette (multiple crown) appearance.

2. Roots are usually stunted.

3. In cauliflower, boron deficiency results in small, deformed heads tinged with reddish-brown areas.

4. In root crops, boron deficiency results in brown heart.

5. Fruit crops—especially apples and pears—develop internal corking (brown) or rot.

6. Cracked stem on celery.

7. Heart rot and dry rot in beets.

8. On some crops portions of large healthy leaves may suddenly collapse and die.

Prize-winning curds of both cauliflower and broccoli grow on healthy plants.

Because some crops are boron-sensitive and others are boron-tolerant, it is important to consider crop rotation.

Boron—B

Section 8:
SYMPTOMS OF BORON EXCESS

Symptoms of boron excess are very similar to boron deficiency or to excess nitrogen.

The first sizable encounter America had with boron toxicity was during World War I, when America was forced to develop its own supplies of potash salts to replace those formerly purchased from German producers. The most easily developed source was that at Searle's Lake in California.

When fertilizers containing potash from that source were first used on crops, widespread damage resulted, especially to potatoes and vegetables. Investigation revealed that boron in toxic concentrations was present in these fertilizers.

The amount of boron that plants need is about 25 to 75 ppm. If more is available, many plants will absorb it in toxic amounts.

In respect to plant tolerance, boron is different than some of the other trace elements. The margin between the amount required and tolerance is very low.

The symptoms of boron excess are yellowish-brown spots around the edges and interveinal areas of the leaves, particularly on the older leaves.

Excess boron usually produces a progressive necrosis of the leaf, beginning at the tips and margins of the leaves. The leaves quickly take on a burned or scorched appearance. Later the affected leaves drop prematurely. Severe wilting of the leaves are the first symptoms on very sensitive crops such as beans and strawberries.

In boron deficiency, cracked stems develop on celery stocks and hearts of celery become necrotic.

The crop of celery has no deficiency.

Mr. Champagne is displaying some prize trophies he helped grow on land previously called "devil land" because crops failed to grow.

Boron—B

Section 9:
CAUTIONS REGARDING THE USE OF BORON

Just like other soil elements, boron is toxic to plants when the concentration is high. Such conditions may be natural to the soil, or they may be the result from applications of salts of the various elements in fertilizers, insecticides, or fungicides.

Sewage sludges, as an example, are often very high in some of the trace minerals.

Observe the Following Precautions for Boron:

1. Boron is one of the oldest weed killers known to man and was used for that purpose. The active ingredient is boron.

2. Be aware that the boron content of the soil can be high in low rainfall regions. Also, that the irrigation waters may contain enough boron to injure crops.

3. There are areas in which boron toxicity does exist. If the need arises, practice the following control measures to prevent accumulation of toxic concentrations of boron.

   a. Leach contaminated soils thoroughly. Available boron moves with the soil moisture. Excess boron can be leached out of soils.

   b. Mix irrigation water that is high in boron content with water that is low in boron. This allows using all the water available for irrigation without building up toxic concentrations.

Boron—B

White stocks and perfect leaves on bok choy (Chinese cabbage) has no deficiency.

c. Do not use boron fertilizers indiscretely! The effect may be crop failure.

d. In humid regions boron toxicity is only temporary. The boron is removed from the soil through leaching.

e. As a general recommendation to farmers, apply 10 to 20 pounds sodium borate per acre on experimental plots only until experience shows the need for more boron.

The crop can be produced ready for market in 31 days.

A crop of green swiss chard is free from deficiencies and ready for a good harvest.

Such crops as red swiss chard add color and beauty to a garden of assorted vegetables.

Crisp leaves and stocks of red chard are nutritious and important as blood purifiers.

Boron—B

Section 10:
SPECIAL SUGGESTIONS REGARDING BORON

Just as the housewife knows that plastic has its limits, the farmer knows that the use of boron has its limitations.

A crop can fail for a lack of 5 to 10 pounds of boron per acre. And a crop can fail from a mere excess of 10 pounds boron per acre.

It seldom happens that the available boron in acid soils is high enough to be toxic.

The recommendation is to apply 10 to 20 pounds of sodium borate per acre yearly on acid to neutral soils. This amount should be sufficient to meet boron requirements for most crops and yet not be toxic to boron-sensitive crops such as beans and strawberries.

When boron deficiency occurs, apply 20 pounds of borax per acre as a corrective treatment.

Beware of mixing borated fertilizers with seed at the time of planting, or banding borated fertilizers too close to the seed at the time of planting.

When using borax straight as a corrective treatment, mix the amount with an inert material, such as sand, peat moss, sawdust, rice hulls, etc. This is to increase the volume to simplify the problem of even distribution.

Boric acid is recommended either for foliar sprays or for soil application. Boric acid is compatible with ammonium compounds. The high-grade trioxide borate fertilizers should *not* be used with straight ammonium salts because of a possible chemical reaction.

A crop of healthy eggplant.

Well-shaped fruit of eggplant indicate the plant is healthy.

Some markets demand medium-sized fruit. Others prefer large fruit. There is a fine line between boron deficiency and boron excess. Thus, accuracy in weighing and applying is important.

Boron—B

Boron-sensitive Crops

Most fruit trees, grapes, most beans, some annual flowering plants, strawberries, etc.

Crops That Are Tolerant of Boron

Beets, turnips, cotton, asparagus, alfalfa, clover, etc.

Crops That Are Intermediately Sensitive to Boron

Most grains, peppers, tomatoes, potatoes, cabbages, radishes, lettuce, and carrots.

Boron—B

Section 11:
SUMMARY AND REVIEW OF BORON

Partial List of Boron Fertilizers

Borax (sodium borate)	10.5 percent B
Boric acid	69 percent B
Solubar	20.5 percent B

Boron Deficiency Symptoms on Miscellaneous Crops

Bean	Enlarged cotyledon leaves; death of terminal buds; leaves are an off-green color; flowers and pods fail to grow.
Beet	Rosette terminal buds; leaves die in the crown; roots show heart rot and dry rot; leaves are red, small, and deformed.
Broccoli	Head and leaf margins have rust-brown discoloration; leaves become necrotic; stems of curds are hollow.
Cabbage	Stems are hollow in head region; leaves are brittle; stiff along margins; leaves making the head are unattached.
Cantaloupe	Rosette crown; death of terminal buds; blossom-end rot on fruit.
Carrot	Margins of leaves are yellow; roots split their length and width.
Cauliflower	Similar to cabbage, except curds are rusty color; deformed with bitter taste; stems are hollow.

There are several predominant symptoms of boron deficiency:
a. Death of the terminal bud. The dead portion in the leaf is an early warning symptom that boron is getting low.

b. Enlarged, thickened, leathery-appearing cotyledon (seed) leaves. When a turgid or crisp-green and otherwise healthy leaf suddenly has a portion collapse, wilt, and die, it is because boron supply is low.

c. Rosette development about the terminal bud area. The necrotic portion on the leaf is boron deficiency.
d. Poor cell division.

Boron—B

Celery	Crosswise cracks in petioles; small center leaves turn brown and rot.
Corn	Growing point dies; no silk and no ears.
Cucumber	Enlarged cotyledon leaves; rosette crown; cracked fruit.
Lettuce	Distorted terminal leaves; death of growing point.
Melon	Large, leathery cotyledon leaves; rosette crown; poor fruit-set; deformed blossom end of fruit; blossom-end rot of fruit.
Onion	Leaves are deep bluish-green color; basal leaves develop transverse (crosswise) cracks on upper side; bulbs are poor.
Peas	Yellow or white veins in leaves; terminal buds die; blossoms drop; pods are empty or hollow and poorly filled.
Potato	Older leaves curl upward at margins; bushy appearance of plant; foliage is thick and brittle; older leaf stocks break easily; tubers are hollow or contain black rings.
Radish	Terminal buds die; tubers crack and become hollow; bulbs may not develop.

The leaves on the cucumber plant are normal—no deficiency.

Boron deficiency results in severely crooked and curved fruit, and along the inside of the curve scar tissue seams frequently appear.

A box of field-run cucumbers taken from healthy vines.

Boron—B

Sweet potato	Terminal buds are deformed; young leaves are ill-shaped; older leaves are small and dull appearing; tubers crack and develop a foul-smelling black slime.
Tomato	Cotyledons and true leaves of young plants turn purple; terminal shoots curl upward and die; blossom-end rot on fruit.

Boron deficiency in lettuce results in the older leaves turning chlorotic at the leaf margins first, then gradually die prematurely. Terminal buds frequently die.

There will be no loss of yield in the crop. All essential nutrients were adequate for best growth.

Boron—B

Lettuce plants that have no deficiency.

Multiple terminal buds result from boron deficiency.

The enlarged cotyledon seed leaves and poorly developed terminal bud indicate boron deficiency.

Notice the leathery, thick appearance of the cotyledon leaves and rosette crown on the melon plant. This is boron deficiency.

Two weeks after correcting for boron, a normal terminal bud grew and the new leaves were normal.

Boron—B

The symptoms of boron deficiency in the plant are enlarged, thickened, leathery-looking cotyledon leaves and rosette crown.

The new leaves are normal after boron supply was corrected.

Deformed melons are common when boron supplies are too low.

Plants such as this one give proof that boron deficiency can be corrected.

Boron—B

The watermelon plant had boron deficiency earlier, but it recovered after boron was corrected.

A sample of the quality produce picked fresh from the garden.

Crinkled, twisted, deformed, small leaves result from boron deficiency.

Healthy melon vines do produce a quality fruit.

The young pepper leaves are normal soon after correcting the boron supply.

Harvesting quality produce is no accident. The grower observes and responds to the needs of the crop.

A colorful bouquet of crisp bulbs to tantalize the appetite.

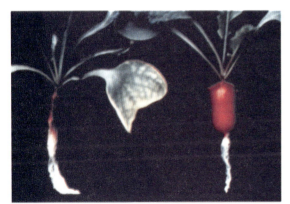

The radish plant on the right is normal. The plant on the left has boron deficiency.

Boron—B

When 5 percent to 10 percent of the radish bulbs split, it is warning that boron supply is low.

The large cotyledon leaves and ill-shaped leaf in the terminal bud are boron deficiency symptoms.

Boron — B

The rosette crown and poorly developed small leaves indicate boron deficiency.

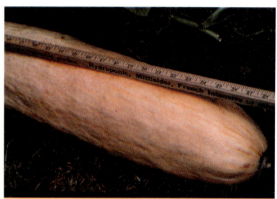

There are many kinds of squash to select from. Some varieties are more tolerant to varying levels of boron than are others.

Healthy squash leaves are large and free from blemishes.

Keeping a close watch on the plants will pay off in crop dividends.

Boron—B

There are no deficiencies on these sweet potatoes.

As the deficiency increases, necrotic and gray blotches develop in the leaflets.

Early boron deficiency symptoms on tomatoes are discoloration and bleaching in the leaflets.

Early symptoms of boron deficiency appear on the terminal buds.

Boron — B

Boron deficiency symptoms appear along the stem of the leaflets as a yellowing. The chlorosis soon extends into the leaflets and advances to the leaf tips.

There are no deficiencies on these staked tomato plants.

Eventually, the entire leaflets are affected, then follows necrosis.

Boron is essential for fruit quality. If it is lacking during the fruiting period, blossom-end rot can develop on the fruit.

Boron — B

Boron is essential for flower development,

. . . and for fruit-set.

MOLYBDENUM
Mo
A Micronutrient

The narrow, long, slightly twisted cabbage leaves from molybdenum deficiency are the reason for the name "whiptail."

The photo illustrates the important role molybdenum performs in plant growth.

Another example of poor plant growth when molybdenum supplies are too low.

Section 1:
DEFINITION OF MOLYBDENUM

The chemical symbol is Mo.

Molybdenum is a silvery-white, metallic element.

It is almost infusible—that is, it is almost incapable of being fused or melted.

The use of molybdenum as a fertilizer in growing crops is increasing so rapidly that it may soon be recognized as a major micronutrient.

The nutrient is vitally important, but only minute applications of the element are required. In fact, the application rate is in grams or ounces rather than pounds.

Two crop diseases stem from molybdenum deficiency:

1. Whiptail of cauliflower, broccoli, cabbage, etc.
2. Yellow spot of citrus.

Molybdenum—Mo

Section 2:
CHEMISTRY OF MOLYBDENUM IN PLANTS

It is generally accepted that molybdenum serves as a catalyst inside plants in their enzyme system.

The enzyme systems reduce nitrate nitrogen to ammonium, which is used in the synthesis of amino acids and protein.

Molybdenum functions in nitrogen fixation by the rhizobia bacteria-producing nodules on roots of leguminous plants and also the nonsymbiotic bacteria (symbiosis is the union or living together of two unlike organisms).

As far as has been established, molybdenum toxicity in plants is in relation to livestock production only.

Toxicity can be experienced in concentrations above 10 ppm (dry weight) in green forage.

Areas of severe molybdenum deficiencies are few, but they do exist—some in California, Florida, Africa, and Australia.

The normal functioning of the life processes of the microorganisms of both green plants and animals depend on available molybdenum.

Applying molybdenum salts to pasture crops must be done with extreme caution.

Molybdenosis disease affects cattle grazing on pastures containing toxic levels of molybdenum. This disease can be fatal to cattle, but it need not occur.

Within plants, molybdenum plays a crucial role in the nitrogen transformation processes.

Following is a list of some of the more sensitive crops to molybdenum deficiency:

An early symptom of molybdenum deficiency is the missing parts along the edge of the cabbage leaf.

The holes and missing parts in the cabbage leaves suggest molybdenum deficiency.

Molybdenum—Mo

The narrow bottom leaves (both left and right sides) of the cabbage plant indicate molybdenum deficiency.

The narrow curled leaves on the red cabbage is whiptail disease, caused from molybdenum deficiency.

The cabbage plants have no deficiency. Notice all the leaves are fully developed.

tomato, potato, lettuce, spinach, celery, beet, all brassicas, and rape.

The characteristic symptoms of deficiency in the brassica group are: narrow leaves with a slight twisting effect called whiptail disease.

In legume crops, the symptoms of deficiency are not as specific. They generally resemble nitrogen deficiency.

It is rare that grasses exhibit distinct molybdenum deficiency symptoms.

As a general statement, the most obvious symptoms of molybdenum deficiency in the early stages are similar to the symptoms of nitrogen deficiency in all crops except grasses and the brassicas.

The predominant symptom of molybdenum deficiency is whiptail disease in cauliflower, cracked stem of celery, etc.

Many growers do not recognize molybdenum deficiency symptoms.

The outstanding feature of molybdenum in crop production is the very small amount required to correct a deficiency.

Applications of the nutrient to correct a deficiency are made in grams or ounces.

To the untrained eye, the deficiency is often mistakenly diagnosed as insect damage.

Sometimes just rolling the seed in molybdenum powder is sufficient to satisfy crop needs.

Molybdenum is essential for crops to perform their proper functions.

Crops such as vegetables and legumes respond fast and favorably to applications of molybdenum.

Twelve ounces (360 grams) of molybdic acid in 55 gallons (U.S.) of water is sufficient to treat more than 1,000 cabbage plants to correct molybdenum deficiency.

Here is a typical example of shredded leaf edges caused from molybdenum deficiency.

Another example of shredded leaf edges caused from molybdenum deficiency.

Molybdenum—Mo

One foliar feeding with molybdenum salts is usually sufficient to satisfy a crop, and one application may last for several succeeding crops.

Molybdenum toxicity has never been recorded as having adverse effects on people.

Molybdenum toxicity has been recorded affecting ruminants (grazing animals).

Fertilizing pastureland with 200 pounds copper sulfate per acre will neutralize the toxic effect of molybdenum.

Another method used to neutralize molybdenum toxicity in sick grazing animals is to inject copper sulfate directly in the bloodstream. The disease is called molybdenosis.

Whiptail disease can be corrected easily with molybdenum salts, providing the proper amount is used and the correction applied promptly.

The production of nitrogen nodules on leguminous plant roots is dependent on molybdenum supply.

A delicate balance seems to exist within the plant between nitrogen, iron, manganese, and molybdenum. All appear to be interdependent on each other.

Molybdenum deficiency affects the manufacture of chlorophyll within the cells of green plants.

Molybdenum deficiencies are nearly as common today as are deficiencies in nitrogen.

Because molybdenum deficiencies are so common today, the symptoms should be better recognized.

Healthy broccoli plants must have adequate molybdenum.

Severe molybdenum deficiency is seen in the long narrow cauliflower leaves.

Molybdenum—Mo

Section 3:
CHEMISTRY OF MOLYBDENUM IN THE SOIL

Native supplies of molybdenum in the soil are very small.

There is usually only one to three ppm, which is less than six pounds an acre in the first eight inches of soil.

Molybdenum toxicity strong enough to affect cattle is less than 10 ppm an acre, or about 20 pounds.

Molybdenum availability to plants is highest when the soil pH is nearly neutral.

In this respect, molybdenum is unlike most of the other essential trace minerals, which are more easily available to plants in acid soils.

Applying sulfuric acid to the soil as a treatment can reduce the availability of molybdenum mainly because it lowers the pH.

The availability of molybdenum is influenced by the following factors:

    1. The pH of the soil, presence of sulfates in the soil, soil phosphates, manganese, and also because of the very small reserves of molybdenum normally in the soil.

    2. Applying lime to acid soils usually increases the availability of molybdenum.

    3. Through ion exchange, fixed (unavailable) molybdate can be displaced by phosphates.

This chemical reaction indicates that the same soil-fixation compounds may be involved in changing both molybdenum and phosphate into unavailable compounds.

After correction for molybdenum deficiency, the cauliflower plants have no deficiency.

Molybdenum — Mo

Molybdenum deficiencies occur more often in acid soils than on neutral soils.

One and one-half pounds molybdenum per acre is an average application.

The maximum application of molybdenum is not more than six pounds per acre.

The total native supply of molybdenum in mineral soils is very small. And like phosphorus, the molybdenum is fixed (unavailable).

Liming acid soils helps to increase the supply of molybdenum for plant use.

Because of the very low native supply of molybdenum in soils, soil tests are not effective in analyzing for molybdenum and therefore are *not* recommended.

Rhizobia bacteria in the soil cannot perform their valuable functions for legume crops without molybdenum.

Brassica crops (cabbage family) are especially sensitive to molybdenum deficiency. If not corrected, entire crops such as cauliflower can fail completely.

Molybdenum—Mo

Section 4:
FORMS OF MOLYBDENUM UTILIZED BY PLANTS

Following is a list of the most frequently used molybdenum fertilizer compounds:

1. Molybdic acid (powdered or liquid)
2. Sodium molybdate
3. Ammonium molybdate

In addition, for special cases, molybdenized phosphate or molybdenized calcium are used to supply molybdenum.

Regular commercial fertilizers seldom contain more than a few parts per million of this element.

Molybdenum fertilizers are effective in foliar feeding and are recommended for this purpose.

The disfigured cauliflower leaves indicate severe molybdenum deficiency.

This brassica crop has no deficiency.

The partial leaf (bottom right) is the result from molybdenum deficiency.

Molybdenum — Mo

Snow-white, well-filled heads of cauliflower have no deficiency.

Section 5:
ACTIONS AFFECTING THE SUPPLY OF MOLYBDENUM

Deficiencies of molybdenum in crops can originate from two sources:

1. The year-by-year weathering of the soil removes the total supply of the element to very low amounts.

2. Acid soils may contain a good total supply, but it may be fixed (unavailable) in forms that plants and soil microorganisms cannot utilize.

Liming acid soils lifts the soil pH, and this frequently releases adequate molybdenum for crop and microorganism needs.

It frequently occurs that the soils that are low in molybdenum are low also in copper.

As an ion, when the pH is below 6.0, molybdenum is tightly adsorbed to the soil colloids and minerals. Ions carry negative charges of electricity.

Heavy intensive farming of soils that have a low native supply of molybdenum and are neutral to alkaline pH values may deplete the available molybdenum.

Increasing the supply of phosphorus improves the assimilation of molybdenum by plants.

Using sulfate fertilizer materials may increase the problem of molybdenum deficiency where both sulfur and molybdenum supplies are low. This may be because molybdenum is required in greater amounts because of the increased growth, which the sulfate fertilizer produced.

Sometimes a few grams of molybdenum is all that is needed to produce prize-winning heads of cauliflower.

Molybdenum — Mo

There are reports that indicate manganese can intensify (bring forward) molybdenum deficiency.

Prolonged use and heavy applications of ammonium sulfate fertilizer may intensify molybdenum deficiency by increasing the acidity of the soil and because the sulfate reduces the absorption of molybdenum.

In some soils with a pH below 5.0, molybdenum becomes deficient.

Liming these soils increases the availability of phosphorus and molybdenum.

The citrus leaf on the right is normal. The three to the left show molybdenum deficiency.

The missing portion in the leaf and crinkled center portion on the cucumber plant are signs of molybdenum deficiency.

These perfect cucumber leaves have no deficiency.

Molybdenum—Mo

The poorly shaped or missing tissue in the melon leaves is molybdenum deficiency.

The deformed leaves on the melon plant are from molybdenum deficiency.

The missing portions in the leaves are typical for molybdenum deficiency.

Section 6:
METHODS OF APPLYING MOLYBDENUM TO PLANTS

The usual method is to take a soluble molybdenum compound, such as sodium or ammonium molybdate or molybdic acid powder, and mix it with superphosphate or other mixed fertilizers and apply all together to crops.

Sometimes these are broadcast on the soil before seedbed preparation.

Sometimes they are banded at the time of planting seed.

For drill-seeded crops such as soybeans, bushbeans, alfalfa, melons, corn, peas, etc., the soluble molybdenum powder can be mixed directly with the seed along with the rhizobia bacteria powder in the seedbox at the time of planting.

Another method is to mix soluble molybdenum powder with water and apply as a foliar spray on crops.

Frequently, phosphate, sulfate, and molybdenum deficiencies occur at the same time.

This field condition prompted the manufacture of molybdenized superphosphate.

As the name implies, this product contains all three elements listed above and is used to correct all three deficiencies in one fertilizer compound.

Because molybdenum requirements and application rates are so very small, it is necessary to mix it with other materials to increase the volume so it can be applied evenly and accurately on the soil or to the crop.

Molybdenum—Mo

Sometime molybdenum is mixed with slightly damp peat moss, sawdust, perlite, or sand before it is applied to the crop or field.

Whiptail disease of cauliflower and other brassica crops can be cured if the proper application of molybdenum is dissolved in water and applied as a foliar spray directly on the crop.

To be most effective, the treatment must be applied promptly. Early detection of deficiency symptoms is important.

Frequently, the only early symptom revealing molybdenum deficiency is missing portions along the leaf edges or sometimes in the leaves.

Notice the difference! When insects eat plant leaves, the damaged areas have sharp edges. In molybdenum deficiency the edges are rounded—not sharp.

Molybdenum deficiency affects the manufacture of chlorophyll within the cells of green plants.

Molybdenum—Mo

Section 7:
SYMPTOMS OF MOLYBDENUM DEFICIENCY

The predominant symptoms of molybdenum deficiency are whiptail disease on cauliflower and other brassica crops; cracked stem of celery, etc.

Whiptail disease in cauliflower develops characteristic narrow, twisted leaves with irregular edges.

The symptoms of acute deficiency are death of small areas of leaf tissue between the veins; the leaves lengthen abnormally and twist slightly.

In mild deficiency, the cauliflower heads do not develop. This can occur even though the deficiency is too mild to produce whiptail disease in the leaves.

If corrective treatment is applied promptly, whiptail disease can nearly always be corrected.

The practice to routinely apply one pound or more per acre of soluble molybdenum salts is becoming an accepted precautionary procedure by many vegetable growers.

Molybdenum is essential for nitrogen utilization in plants.

Molybdenum-deficient plants are stunted and have varying shades of yellow discoloration closely resembling nitrogen deficiency.

A very common deficiency symptom is missing parts of leaves, either on the edges or in the tissues in the leaves; or it can resemble scar tissue on leaves.

Severe molybdenum deficiency is seen on the melon leaves.

Notice the difference a ready supply of molybdenum has on the leaves and plant growth.

The old melon leaf (bottom) has molybdenum deficiency. The new leaf (center) is normal after the deficiency was corrected.

This melon field recovered completely from molybdenum deficiency.

Molybdenum—Mo

Section 8:
SYMPTOMS OF MOLYBDENUM EXCESS

Molybdenum toxicity or excess is rare and seldom recognized in the field.

Plants seem to tolerate high concentrations of this element in their tissues.

For livestock (ruminant animals), molybdenum excess can be especially serious.

Molybdenum toxicity is known by several names:

1. Molybdenosis
2. Teart disease
3. Peat scows

Encounters with ruminant animals having molybdenum toxicity have been from naturally occurring excess molybdenum, either in the soil or in the irrigation water.

Treating forage crops with copper sulfate has been successful in counteracting the harmful effects of molybdenum toxicity in ruminant animals.

Plants may accumulate 100 ppm or more of molybdenum without noticeable symptoms of toxicity to the plant itself.

Molybdenosis disease in livestock may be encountered on forage crops that contain less than 10 ppm of molybdenum in their tissue.

Molybdenum—Mo

Section 9:
CAUTIONS REGARDING THE USE OF MOLYBDENUM

Nearly all the toxic elements in the soil may occur naturally, or they may result from applying salts of the various elements in fertilizers, insecticides, or fungicides.

Under certain conditions of toxicity, indicator plants thrive and reveal the toxic elements.

In the case of molybdenum, cattle grazing on the forage crops may be poisoned.

Molybdenum is unique among all the essential trace elements in the very small amount required. Applications are always figured in grams or ounces instead of pounds.

For many soils that are low in this element, fertilizing with one to one and one-half pounds to the acre is liberal and frequently one application will satisfy crop needs for two or three years.

General field applications of molybdenum, however, are usually 1 to 4 pounds per acre.

Rhizobia bacteria are very dependent on molybdenum to perform their normal functions in legume plants.

Molybdenum is essential to the processes of nitrogen transformation in plants.

Phosphorus and sulfur are often deficient under the same conditions that bring about molybdenum deficiency.

But increasing the supply of molybdenum through fertilizers will produce only negative crop response until the phosphorus or sulfate fertility levels are increased adequately.

Well-shaped melon leaves with dark-green color and strong vines indicate to the grower that the essential nutrients were adequate.

Blue-green color on leaves, dying older leaves, and poor growth of the onions are symptoms of molybdenum deficiency.

Molybdenum — Mo

Section 10:
SUGGESTIONS REGARDING MOLYBDENUM

Molybdenum materials are very expensive.

Prices range from $2.90 to $55.00 per pound. The average price being around $15.00 per pound.

For small greenhouse or family garden, the recommended molybdenum fertilizer application is four grams per each 25 pounds of mixed fertilizers.

In whole numbers, about 30 grams is equal to one ounce, and 16 ounces equals 480 grams.

The use of molybdenum in growing family gardens is not widely practiced, but is increasing rapidly.

The narrow, long radish leaf is a symptom of molybdenum deficiency.

The missing tissues in the squash leaf indicate molybdenum deficiency.

The melon field has no deficiency.

The missing parts in the tomato leaves indicate molybdenum deficiency.

Notice the narrow, ill-shaped leaves? This is a typical pattern in molybdenum deficiency.

Molybdenum—Mo

Section 11:
SUMMARY AND REVIEW OF MOLYBDENUM

Partial List of Molybdenum Fertilizer Materials

Sodium molybdate

Ammonium molybdate

Molybdic acid

Effects of Molybdenum Deficiency on Miscellaneous Crops

Bean	Leaves are pale-green color with interveinal mottling; brown scorched areas develop rapidly in the interveinal tissue; green bands remain close to the middle ribs and veins, even after death of affected tissue.
Beet	Blades are pale-green and curl upward from the middle rib; red veins are conspicuous because of chlorotic leaves.
Broccoli	Leaves develop whiptail appearance; necrosis develops in the interveinal areas; cotyledon leaves usually remain green.
Brussels sprouts	Leaves become cupped; young leaves become twisted; plants are gray-green color.
Cabbage	Older leaves are mottled, scorched, bleached, and cupped; heart formation is poor.

Molybdenum—Mo

The small, narrow tomato leaflets (right) indicate molybdenum deficiency. The perfect leaves (left) have no deficiency.

Another symptom of molybdenum deficiency is chlorosis (bleaching) of entire leaflets and leaves, including the leaf stems.

The symptoms on the leaf indicate severe molybdenum deficiency.

Cantaloupe	Color of leaves is pale green; Yellow-green chlorosis on the interveinal tissue; brown marginal scorching is followed by severe marginal withering.
Carrot	Older foliage becomes pale-green color with leaf-tip scorching
Cauliflower	Whiptail disease; leaves are twisted and elongated; thin (laminae) membrane are cupped, relatively turgid (stiff), and very dark-green color.
Celery	Foliage becomes uniformly bright-yellow color, without interveinal mottling; portions of older leaves scorch.
Chard	Leaves are pale yellow; margins cup and become necrotic.
Cucumber	Yellow patches develop on older leaves; these become scorched; margins roll upward.
Lettuce	Leaves are pale green; margins cup and necrose; leaves may wilt and scorch; hearts fail to develop.
Peas	Plants are pale yellow; older leaves develop mottling and die quickly; terminal leaves remain green; flowers and seed production are poor.

Molybdenum — Mo

The bleached-gray necrotic condition of the tomato leaf is a symptom of molybdenum deficiency.

The necrotic condition of the squash leaves and the crinkled younger leaves indicate molybdenum deficiency.

Potato Leaves are first pale, then later develop a bright-yellow chlorosis; chlorotic areas are irregular; leaf margins are frequently darker green; leaves curl upward; young leaf and flower buds turn brown and wither before opening, followed by death of the growing point.

Radish Bright-yellow interveinal mottling (spotting); cotyledons are large and green; margins of leaves cup upward and become necrotic.

Spinach Leaves are pale yellow and cupped, with severe chlorosis and necrosis.

Tomato Leaves have interveinal yellow mottling; margins cup upward, and leaflets appear rolled; in severe cases plants die; in mild cases flowering and fruiting are affected.

Molybdenum — Mo

The living-green color of the perfect leaves and many large flowers indicate the supply of essential nutrients was adequate.

Healthy tomato plants do not drop their flowers or fruit.

Ripe tomatoes are always popular, and their tantalizing flavor stimulates the appetite.

SALINITY
Soil pH
Symptoms of Distress

The white color along the ridges of the soil beds is called salinity. It indicates an accumulation of certain salts (chemical compounds) in the soil and soil surface.

The pH of the soil is between 7.8 and 8.3. The salinity is high, but good crops were produced repeatedly.

Too much fertilizer applied as a top-dressing too close to the plants destroyed the crop.

Salinity Symptoms of Distress

The pH (salinity and alkalinity) of the soil is important in the respect of selecting the right fertilizer compounds and the availability of the applied fertilizers to growing crops. Many saline soils can be reclaimed to produce high-yield crops.

Salinity—Soil pH

The dying and dead zucchini plants are the result of incorporating too much fertilizer into the soil before transplanting the plants.

The poor seed germination was a result from too much fertilizer.

The bare area is the result of applying too much dry fertilizer as a top-dressing at the time seeds were germinating.

Excess fertilizer not only retards seed germination but can also kill the seed.

The bean leaves are suffering from excess boron fertilizer.

Strawberries and beans are very sensitive to excess boron. The damage to the leaves is excess boron.

This type of damage to bean leaves indicates excess boron.

Salinity—Soil pH

The abnormally yellow leaves and salt accumulation seem to indicate manganese deficiency; in reality, it is high excess salt.

This type of damage on the leaf edges is typical of excess fertilizer.

Salinity—Soil pH

The bleached leaf edges indicate high salt in the root area.

The leaves and the fruit both indicate excess fertilizer.

Fruit-set, if any, is very poor when salt accumulation affects the leaves.

The bleached borders of the leaves result from too much fertilizer in the root area.

Salinity—Soil pH

High salt concentration about the roots will result in abnormal wilting of plants, even on cool and cloudy days.

There is evidence in the photo of excess fertilizer.

The symptoms on the pansy leaves are a result of high excess salt (fertilizer).

Dry necrotic leaf edges are symptoms of mild excess fertilizer. Heavy leaching saved the crop.

Salinity—Soil pH

The uneven, stunted growth of the pansies resulted from too much fertilizer. Prompt heavy leaching saved the crop.

Early symptoms of excess fertilizer are evident on the leaf edges on this tomato plant.

Too much fertilizer and poor drainage destroyed this tomato crop.

High salt and poor drainage can destroy a healthy crop during any period of growth.

TWO OR MORE DEFICIENCIES

Previously each of the 13 essential plant nutrients that man can regulate (carbon, hydrogen, and oxygen are usually not considered) have been considered separately, and illustrations of the various single deficiency symptoms have been given.

In actual practice, however, it is more common to experience problems involving two or more deficiencies at the same time. Thus, the following deals with crops indicating more than one deficiency symptom.

The pale-green-colored cabbage head indicates nitrogen deficiency. The purple-colored older leaves indicate phosphorus deficiency.

The blue-green color of the cabbage head and necrosis of the leaf edges indicate nitrogen and potassium deficiency.

The uniform orange-yellow color, including the leaf veins, indicate nitrogen and magnesium deficiency.

Small sweet potato leaves and dull-bronze color indicate nitrogen deficiency. The purple discoloration suggests possible phosphorus deficiency.

Two or More Deficiencies

Crops need not fail. They can be healthy if properly and adequately fertilized.

Lyman is observing the effects nutrient deficiencies had on recently planted soybean seed. No fertilizers were applied to the poorly stunted, discolored row of soybeans when planted.

The size and shape of the sweet potatoes indicate that nutrient supply was very low.

Two or More Deficiencies

Before a successful crop of soybeans could be grown, nitrogen and boron had to be applied on the land.

Purple-colored leaves are symptoms of phosphorus deficiency. The missing parts on the leaves indicate potassium deficiency.

Purple color on the topside of the oldest bean leaves indicate phosphorus deficiency. The dead and dying old leaves indicate calcium deficiency.

Purple blotches on top of the older leaves and purple blush indicate phosphorus deficiency. Missing parts of leaves is molybdenum deficiency.

Two or More Deficiencies

The purple on the leaves suggests phosphorus deficiency. The yellow color on young leaves is manganese deficiency.

Purple color on rim of broccoli leaves suggests phosphorus deficiency. Light-yellow color overall indicates manganese deficiency.

Purple color indicates phosphorus deficiency. Torn leaves and gray spots indicate magnesium deficiency.

Purple blush on the pepper fruit suggests phosphorus deficiency. The light-colored areas in the older leaves is magnesium deficiency.

Two or More Deficiencies

Reddish color on the leaf suggests magnesium deficiency.

Red leaves on bottom of the corn plant is magnesium deficiency.
Purplish color on top of leaves is phosphorus deficiency.

Light-green areas in the leaves is magnesium deficiency.

Light-green blotches in the bean leaves is magnesium deficiency. Necrotic-brown spots in the leaf-edges is potassium deficiency.

Two or More Deficiencies

Light-green-colored areas in the bean leaves is magnesium deficiency. Torn leaf edges is potassium deficiency.

Bright colors in the small carrot leaves and deteriorating older leaves indicate magnesium and potassium deficiencies.

Stunted growth, small beet leaves, and necrotic older leaves suggest both potassium and magnesium deficiencies.

Bright colors on the carrot leaves and necrotic older leaves suggest magnesium and potassium deficiencies.

The scar seam along the curve on the bean pods is boron deficiency. Restricted growth and empty pods along the stem end indicate potassium deficiency.

Two or More Deficiencies

A collapsed dead portion of the leaf is boron deficiency.

The gray dead area in the cucumber leaf is boron deficiency. The gray spots in the leaf is potassium deficiency.

An orange-yellow color along the leaf borders is magnesium deficiency.

Two or More Deficiencies

Large leathery cotyledon leaves indicate boron deficiency.

Large, ill-shaped cotyledon leaves indicate boron deficiency. The upturned and rough leaves indicate potassium deficiency.

Deformed leaves indicate boron deficiency. Holes in the leaves and edges indicate potassium deficiency.

Large necrotic cotyledon leaves indicate boron deficiency.

Yellowish color and a warty appearance of the leaves indicate calcium deficiency.

Two or More Deficiencies

Warty appearance on some leaves is calcium deficiency. Injured leaf edges indicate potassium deficiency.

Warty appearance of the leaves is calcium deficiency. Torn leaves suggest potassium deficiency.

Same as above.

Two or More Deficiencies

Raised rough areas in the leaves indicate calcium deficiency. Necrotic brown areas indicate potassium deficiency.

Missing portions on the leaf indicates molybdenum deficiency. Rough leaf borders suggest potassium deficiency.

Yellow-brown specks on older leaves suggest zinc deficiency. Necrotic and torn leaves indicate potassium deficiency.

Warty leaves indicate calcium deficiency. Rough small leaves suggest potassium deficiency. Healthy leaves on the left are new leaves after correction.

Two or More Deficiencies

Uniform pale-yellow-colored leaves suggest magnesium deficiency.
Necrotic dry leaf edges indicate potassium deficiency.

The three ears on the left are normal.
The center ear is deficient in phosphorus.
The fourth ear from the right is deficient in nitrogen.
The second ear from the right is deficient in boron and phosphorus. The first ear on the right is deficient in nitrogen.

Six ears of corn on the left are normal.
The other three ears are deficient in nitrogen, potassium, and boron.

The yellow coloring in the interveinal tissue is zinc deficiency. Necrosis on the leaves is potassium deficiency.

Two or More Deficiencies

The orange-colored section of old leaf is magnesium deficiency. Death of older leaves indicates calcium deficiency.

Three weeks after correction for zinc, manganese, and potassium, the crop was healthy.

The orange-colored section of old leaf is magnesium deficiency. Death of older leaves indicates calcium deficiency.

Light-green areas in the leaves suggest magnesium deficiency. Necrosis of leaf edges is calcium deficiency.

Two or More Deficiencies

Deformed leaves indicate boron deficiency.
Dying and drying leaves indicate calcium deficiency.

Ill-shaped, deformed small leaves suggest both calcium and boron deficiencies.

The ruffled green leaves and upturned leaf edges suggest both boron and calcium deficiencies.

The downward rolled leaves and missing leaf parts indicate molybdenum deficiency.
The upward curled leaves and deformed leaves indicate calcium deficiency.

Light-green patches in whole leaves suggest magnesium deficiency. Bright-yellow to orange leaves suggest zinc deficiency.

Two or More Deficiencies

The curled, dwarfed leaves and missing leaf portions suggest boron and molybdenum deficiencies.

Deformed, curled leaves and missing sections indicate boron and molybdenum deficiencies.

The melon leaves indicate boron and molybdenum deficiencies.

Two or More Deficiencies

The yellow to orange coloration on the corn leaves indicates both zinc and manganese deficiencies.

This photo shows zinc and manganese deficiencies on squash leaves.

Because both the veins and interveinal tissue are the same color, the problems are zinc and manganese deficiencies.

The yellow-colored leaves in the framed bed are deficient in zinc and manganese.

Two or More Deficiencies

Light-yellow-colored leaves indicate manganese deficiency. The green veins indicate zinc —and possibly iron—deficiency.

Young plants were slowly dying because of zinc and manganese deficiencies.

The plant was dying because of deficiencies in zinc and manganese.

Not all plants in the same variety are equally sensitive to zinc and manganese deficiencies.

Two or More Deficiencies

To save the crop, two applications of manganese and zinc had to be applied.

Both small corn plants have severe deficiencies in zinc and manganese.

If only zinc was deficient, there would be ribbons of yellow on the leaves. When entire leaves are yellow, both zinc and manganese may be required.

The same crop as shown above after correcting the deficiencies.

Two or More Deficiencies

The beans have recovered after correcting zinc and manganese deficiencies.

The photo shows deficiencies can be corrected without sustaining a loss in yield or quality.

Notice that all the plants have recovered completely from zinc and manganese deficiencies.

While small, the crops indicated zinc and manganese deficiencies.

Two or More Deficiencies

The photo shows the quality of the crop after correcting the early deficiencies.

The leaves are normal.
They recovered completely from zinc and manganese deficiencies.

All the crops recovered from zinc and manganese deficiencies after the supply was increased.

The squash leaves are recovering from zinc and manganese deficiencies.

Two or More Deficiencies

The yellow-colored areas in the cabbage leaf indicate manganese deficiency. The bleached-white areas suggest sulfur deficiency.

The land had to be fortified with nitrogen, phosphorus, potassium, magnesium, calcium, boron, zinc, and manganese before crops could be grown successfully.

Originally, nitrogen, phosphorus, potassium, magnesium, calcium, and boron had to be applied to the soil before crops could be grown successfully.

High-yield and quality crops can be grown on very poor land when it is fertilized properly.

Two or More Deficiencies

The field of sweet potatoes is healthy because nitrogen, phosphorus, potassium, magnesium, calcium, and boron were applied to the land before planting.

Crops failed because the land was infertile and infested with soil-inhabiting insects.

All crops failed because the land was depleted of calcium, boron, nitrogen, phosphorus, potassium, and magnesium.

Discolored plants indicate real stress problems and will fail to mature.

Crops failed for the same reasons given above.

Two or More Deficiencies

Crops fail on land depleted of essential nutrients, especially when weeds are allowed to compete.

Nearly all crops require the same nutrients. The soil was very low in plant food.

When crops fail on good land, the reason is depleted essential nutrients.

The small sweet potatoes are 22 months old. Had the soil been properly fertilized, a crop could have been harvested in 7 months.

Two or More Deficiencies

Soybeans failed for want of fertilizers and poor weed control.

Later, after fertilizers were applied, soybeans grew well.

All the crops grew well because the land was fertilized. The land is the same as shown lower right, page 96.

Four weeks previously, the crop was dying because fertilizers were lacking.

Two or More Deficiencies

Hungry plants are easy to recognize.

Notice the uniform growth of crops growing in fertile soils.

Five essential nutrients had to be applied to the land before successful crops could be grown.

Cotton leaves show the importance of proper fertilizing practices.

Two or More Deficiencies

The crop was dying. Six essential fertilizers were applied in a narrow band beside the row of plants. After being dissolved through irrigation, the fertilizers were available and the crop recovered.

The photo shows the same variety of sweet potato as shown at lower left, grown on the same land, which was fertilized adequately. The crop matured in 7 months, compared to 10 months in previous attempts.

The photo was taken 6 weeks after fertilizers were applied. Same field as above.

The small, crooked sweet potatoes grew for 10 months in infertile, good-textured soil.

7 RECOGNIZING PLANT DISTRESS SYMPTOMS

You can learn to recognize the symptoms of plant distress.

The stunted growth and small cotton balls are the result of low soil fertility.

MISCELLANEOUS SYMPTOMS

The ground is very rocky. But where water is available and fertilizers are used, good crops can be grown.

Alkaline soils can be made productive by feeding with a balanced nutrient formula and providing adequate drainage.

Miscellaneous Symptoms

The uneven growth and small, poorly developed cabbage leaves indicate infertile soil.

Another soil problem is soil maggots. They can destroy entire crops.

The swellings on the roots are from eelworm (nematodes) infestation. As the eelworms multiply, the damage to the crop increases.

Healthy white roots indicate to the grower that all is well.

Miscellaneous Symptoms

Grasshoppers can multiply rapidly and ruin entire crops.

Soil maggots attacked this crop of beans as the new plants were breaking through the ground. The crop was ruined.

Green cabbage worms spoil the heads of cabbage and limit the formations of the leaves.

Aphids multiply so rapidly they can cover the entire plant. They feed on plant sap and cause severe damage to leaves, flowers, and fruit.

Miscellaneous Symptoms

Leaf miners can destroy the leaves and thereby ruin the harvest.

Lack of essential fertilizers starved and ruined these tomato plants.

The bare ground in the low areas is where water stood during rainy weather and the plants drowned.

The poor bean plants in the low areas drowned from water standing for only a few hours.

Miscellaneous Symptoms

Any time a plant wilts, it has already stopped growing. Therefore, do not allow crops to wilt.

Weeds are robbers of valuable nutrients. Destroy them while very small.

Notice the dead blossoms and no fruit-set. Whenever melon vines are subjected to frequent wilting, they abort flowers and fruit, because of trying to keep the leaves from scorching.

This field had ample moisture, but fertility was low and the weeds were too much competition for the crop. It failed to mature.

The bleached, dying leaves and dull-green and yellow color of the grass indicate low fertility.

Miscellaneous Symptoms

Weeds in a field of sweet potatoes drastically cut the yield, which delayed the harvesting many days.

These tubers came from starved vines after 22 months of growth.

Small leaves and short sweet potato runners are symptoms of low fertility.

Too many plants in one hill greatly reduced the yield mainly because fertility was too low.

Miscellaneous Symptoms

The shape of the sweet potatoes indicate fertilizer deficiencies during the growing period.

The white bands between the rows are fertilizer. However, the plants were starving because the fertilizers were not dissolved and made available to the crop.

Planting carrot seed too thick results in heavy leaf growth, but no bulbs.

Typhoon winds destroyed the crop of zucchini squash.

Miscellaneous Symptoms

Typhoon winds destroyed the crop of bok choy Chinese cabbage.

Whenever a large percent of the tomato fruit cracks at the stem end, it frequently is because of the variety.

Variety is important. The problem with the tomatoes is poor variety.

As it is with people, plants too are affected with too little or too much—even of good things.

Miscellaneous Symptoms

The tomato leaf is sunburned. Such an injury is not serious unless the damage is severe.

The lower leaves are dying from a lack of light. There was too much shade from leaves higher up.

The gray bleached areas on the bottom of the leaf is salt burn from dry fertilizer granules dropped on the leaf.

The black rot on the cabbage leaves and head is a salt burn caused from granules of fertilizer dropped on the leaves next to the head.

Insecticides can reduce yields. The leaf (lower left of photo) was damaged with Diazinon, used as a soil drench to kill soil maggots.

Miscellaneous Symptoms

Pyrethrum insecticide damaged the leaves on the crop of milo.

The large field of milo was sprayed with Pyrethrum. The damage was temporary because only the older leaves were affected.

Bark mulch was used as a soil medium and, as was learned later, contained traces of 2-4-D herbicide. The crop had to be replanted.

Beware, insecticides and herbicides are poison. They can kill!

Miscellaneous Symptoms

Creosote (a wood preservative) is toxic to plants.

Another example of a potato crop when only compost was used. Crops were very poor.

Seldom does compost supply adequate plant food to mature a crop. In this photo only compost was used, and the crop failed.

Still another crop that failed because only compost was used.

Miscellaneous Symptoms

The soybeans lacked several essential nutrients and had to compete with the weeds. The crop failed.

A crop that failed because of low fertility and weeds.

Low fertility and weeds ruined the crop of vegetables.

Likewise, these crops failed because essential fertilizers were not supplied.

Low fertility and eelworms (nematodes) destroyed this garden.

Miscellaneous Symptoms

The six ears on the left have no deficiency. The three smaller ears have potassium, boron, and nitrogen deficiency.

Accuracy is important. Fertilizers should be weighed or measured before applying them.

The method recommended to eliminate crop failure is to add the essential fertilizers to the land before planting a crop.

Lime is usually applied separately because of the fineness of the particles compared to other fertilizer materials.

Miscellaneous Symptoms

They should be spread evenly.

Frequent light applications of essential fertilizers are recommended to feed growing crops.

Careful mixing of the fertilizers with the soil is recommended.

Fertilizers should be weighed and applied evenly.

Radishes require fertilizer promptly after sprouting. Two applications during the first 10 days growth are all that is needed to mature the crop.

Miscellaneous Symptoms

Other symptoms on crops are related to disease. Wilt disease killed the tomato plants in this photo.

Fertilizing is easy, accurate, and fast when using a simple homemade device.

Orange-gray areas in cauliflower and cabbage leaves is bacterial disease in the plant sap.

Miscellaneous Symptoms

The taro bulbs have scab disease. The disease retards growth.

The ill-shaped thin leaves on the zucchini squash are caused by a virus disease in the plant sap.

The long gray-colored heads of wheat are empty. Blight disease during periods of damp weather struck the heads during the flowering and pollination period.

Curley top, a tomato disease, strikes about mid-growth. Insects called thrips frequently carry and spread the disease.

Miscellaneous Symptoms

Wilt disease kills tomato plants at all stages of growth. It is more severe when fruit is developing.

Late blight disease affects all parts of tomato plants, including the fruit. The disease spreads rapidly.

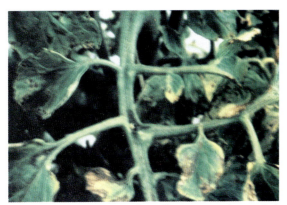

Early blight, a disease that affects tomato leaves but not the fruit.

Late blight develops a foul-smelling, oozing slime on the tomato fruit.

Miscellaneous Symptoms

There are two types of blossom-end rot. Above is the fungus type, which penetrates the fruit and becomes a watery, slimy mass.

Here is the type caused from some type of stress in the plant—either from a lack of water or a nutrient deficiency.

Tobacco mosaic spoils the fruit and the crop.

Tobacco mosaic is another serious tomato disease. It is usually fatal to the crop.

Miscellaneous Symptoms

It affects the fruit in several ways.

In tobacco mosaic even the small fruit is spoiled.

Virus infection in zucchini squash stops the plant from fruiting, and it slowly dies.

Cracked fruit can usually be corrected by changing to a variety that does not crack.

Note: Planting only healthy plants in fertile soil to keep them growing steadily to maturity will reduce almost all adverse problems of production to a minimum.

Garden Grow-Boxes

More Food in Less Space

Grow-boxes are bottomless, wooden or cement frames leveled in place. Usually 5 feet wide, 30 feet long, 8 inches deep; but they can be any size. They can be built almost anywhere. They are filled with "custom-made soil," a mixture of sawdust and sand, or other inert and organic combinations together with a balance of fertilizers.

"Custom-made soil" with its balanced nutrients and proper moisture and feeding produce higher yields and quality on much less space than regular soil.

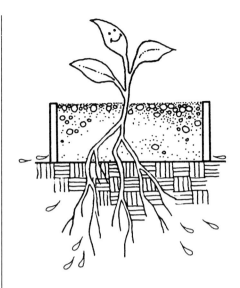

The soft, "custom-made soil" offers many advantages, including perfect drainage, aeration, and balanced feeding. It keeps the usually hard subsoil damp and soft. This lets the roots penetrate the subsoil and absorb many extra important minerals. The growing season is lengthened because it has a cooling effect on roots during the hot summer and because it warms up quickly in the early spring. Level grow-boxes filled with soft, "custom-made soil" save about 40 percent in water; water penetrates uniformly, easily, and quickly.

THE GROW-BOX MUST HAVE GOOD DRAINAGE ALL AROUND THE FRAME.

A special reason for growing plants in grow-boxes is that you can care for the crop from the sides. Shoes carry disease and weed seeds. Do not walk in the grow-boxes!

Grow-box gardening requires few tools. The soft, "custom-made soil" can be worked with basic hand tools or just with the hands.

With grow-boxes you can make gardens of any size. Four grow-boxes can provide a nearly continuous harvest of food much of the year for a family of four, if properly managed. Ten grow-boxes can provide for a family's needs and a large surplus for selling or sharing. Twenty-five to fifty grow-boxes can provide for complete economic self-sufficiency!

EVEN IN AREAS WITH SEVERE WINTER TEMPERATURES, THE GROW-BOXES CAN BE COVERED EASILY TO MAKE INEXPENSIVE GREENHOUSES.

How to Make Grow-Boxes

Select a sunny location. In cool climates, build boxes with a southern exposure near a house, patio, barn or dug into a hillside.

You can even build grow-boxes on poor, hillside land, rocky soil, clay, alkali, or asphalt. It doesn't matter, the grow-boxes will work! Just level or terrace the space for the grow-boxes. Provide good drainage away from the grow-boxes. Roots drown in standing water.

Build the number of grow-boxes needed to fit your plot. Grow-box size can vary to fit unusual boundaries. If several grow-boxes are built side-to-side, line them up straight for attractiveness and easy working. Provide 3'-0" of working space on each side of the box and 5'-0" of working space at the ends of the box.

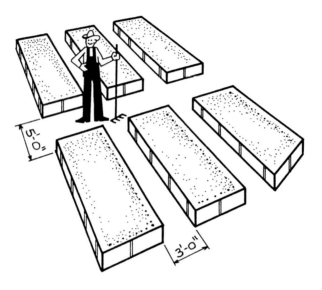

Here is the materials list for the standard 5-foot-wide, 30-foot-long and 8-inch-deep Mittleider grow-box.

A) 70 feet of 1" x 8" redwood or cedar.
B) 24-1" x 2" x 18" long pointed redwood stakes.
C) One pound of 4" nails.
D) One three-pound hammer or mall.
E) One regular claw hammer.
F) 100 feet of strong cord.
G) One level at least two feet long.

HERE IS HOW TO BUILD THE STANDARD 5′x30′x8″ MITTLEIDER GROW-BOX:

Level enough ground for each box area. Establish the location of the corners of the grow-box with the cord. Tie the cord securely to the stakes.

NOTE: Corners should be square (90°) unless the area is irregular in shape.

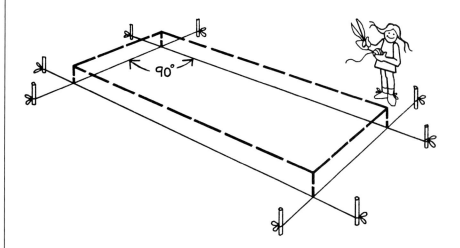

Nine inches from a corner, along one side, drive the first stake into the ground, to depth of about nine inches. Always drive stakes on the outside edges of the grow-box, NEVER on the inside edges. Drive stakes about 30 inches apart along the cord line for one side of the box. When nailed securely to the leveled 1" x 8" boards, the top edges of boards and the stakes should be level (flush).

Nail a 1" x 8" side board flush with top of stake nearest a box end. Drive stake and side board deeper together until bottom edge of side board touched the ground. Proceed to second stake.

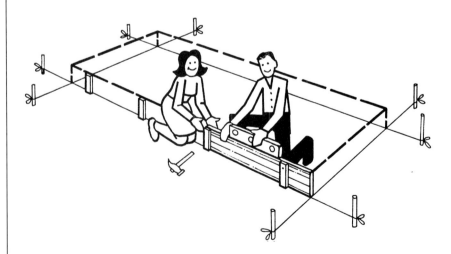

With a level on top edge of board raise or lower board to level position; drive stake deeper until flush with leveled board. Then nail board to stake with two nails. Repeat the same at each stake to other end of box.

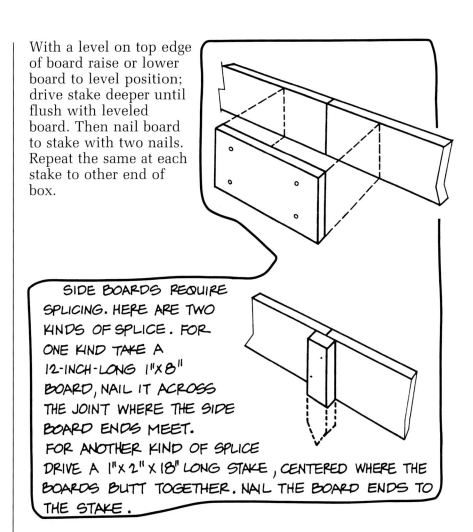

SIDE BOARDS REQUIRE SPLICING. HERE ARE TWO KINDS OF SPLICE. FOR ONE KIND TAKE A 12-INCH-LONG 1"x 8" BOARD, NAIL IT ACROSS THE JOINT WHERE THE SIDE BOARD ENDS MEET. FOR ANOTHER KIND OF SPLICE DRIVE A 1"x 2" x 18" LONG STAKE, CENTERED WHERE THE BOARDS BUTT TOGETHER. NAIL THE BOARD ENDS TO THE STAKE.

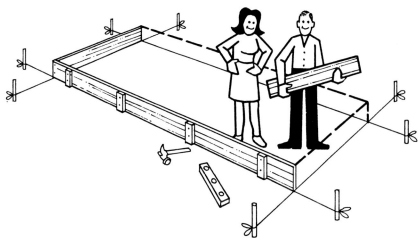

NAIL A 5-FOOT BOX END-PIECE (1"x 8") TO EACH END OF THE LEVELED SIDE.

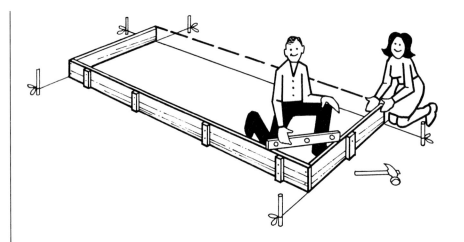

Drive a stake near the center of both 5-foot box ends in line with the cord. Place level like a triangle across tops of both side and end boards. Level the 5-foot board, drive stake to proper depth, and nail to 5-foot end-piece. Repeat the same for opposite end of box.

To level the opposite side board, place level, again like a triangle, across the 5-foot end-piece and the top of the 1" x 8" side board. Level side board to match 5-foot end-piece. Drive stake to proper depth, nail board to stake.

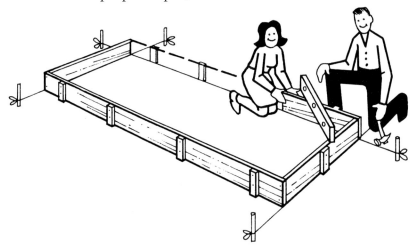

Repeat, as for first side board, and continue to opposite end of box.

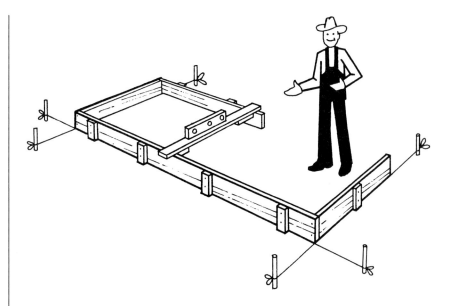

When many grow-boxes are to be built, here is a method to speed up construction. Stretch and stake cords for the two box ends and only one side. Level the side along the cord. Use "spreader board" and the level to level the other side and the two ends.

> **NOTE** "SPREADER BOARD" IS A 2"x4" x 6-FOOT-LONG BOARD WITH 2"x4" x 6"-LONG BLOCKS NAILED 60 INCHES APART (OUTSIDE DIMENSION). THE TWO BLOCKS CONTROL THE INSIDE WIDTH OF THE BOX.

Place the spreader board horizontally across the box. By raising or lowering the loose side boards to the level on the "spreader board" you can make the loose side level, line it up straight, stake, and nail it in one operation!

Using Custom-made Soil

"Custom-made soil" is used to avoid crop failures. Crops fail in stubborn, hard-to-manage soils because of soil disease and insects, gophers, moles, and rabbits. Also there is the constant battle against weeds, and the nutrient deficiencies in soils which are becoming more and more serious. "Custom-made soil"—the heart of the Mittleider method—is used to avoid all these disappointments. It assures success every time, and greatly increases yields!

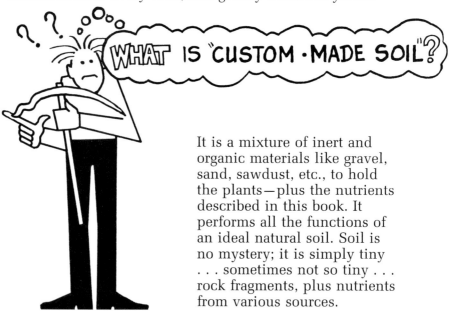

It is a mixture of inert and organic materials like gravel, sand, sawdust, etc., to hold the plants—plus the nutrients described in this book. It performs all the functions of an ideal natural soil. Soil is no mystery; it is simply tiny ... sometimes not so tiny ... rock fragments, plus nutrients from various sources.

Choose any available materials like these and make the combination you like best.

1. 50% blowsand with 50% peat moss.
2. 75% sawdust with 25% fine sand.
3. 50% perlite with 50% peat moss or sawdust.
4. 50% sawdust with 50% styrofoam pellets or pieces.

NOTE: Blowsand is fine sand like that heaped up by the wind (sand dunes).
Perlite is bits of volcanic glass "popped" by heat. It is available at construction suppliers.
Sawdust is safe to use from almost all kinds of wood. Fresh from the saw or aged—either is good.

Remember the soil under your grow-box can be sand, rocks, gravel, good soil, peat, clay, or cement.

Spread ten pounds gypsum (lime) evenly over the inside area of one 5' x 30' grow-box. Fill the grow-box level full with the mixture you chose from the four combinations listed earlier. Do not tamp or pack the soil mixture.

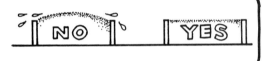

Mix the material thoroughly. A rake or small rototiller will work nicely.

While mixing, add enough water to produce a wet medium, but not so wet you could squeeze water from it.

Here are the ingredients for the Mittleider pre-plant fertilizer mix. Carefully weigh and mix together the following. Usually these materials are available at your garden supply store.

1. 4 pounds double superphosphate.
2. 2 pounds potassium sulfate or chloride.
3. 4 pounds sulfate of ammonia.
4. 2 pounds magnesium sulfate.
5. 2 ounces (60 grams) baron (sodium borate or boric acid).

Spread this dry mixture evenly over the "custom-made soil" in the grow-box. Now spread 5 pounds gypsum (lime) evenly over the grow-box area right on top of the other fertilizers.

NOTE: FOR LIME, USE GYPSUM IN ARID AREAS. USE AGRICULTURAL OR DOLOMITE LIME IN AREAS THAT GET MORE THAN 20 INCHES OF RAIN YEARLY.

Thoroughly mix everything in the grow-box together—the "custom-made soil," lime, and pre-plant fertilizers. Add enough water to make a wet mixture. DO NOT over-fill the grow-box with "soil" mixture. Finish by neatly leveling the complete mixture to the top edge of the grow-box. Sprinkle lightly, using a fine water spray, to keep the soil surface from rapid drying.

Congratulations!!

Your grow-box is complete and ready for planting. You can plant seed directly in the box or you can transplant seedlings from the nursery or from your own starting frames.

Index

Agricultural lime see Lime
Alfalfa
   and borax fertilizers 3:24
   and boron 3:15, 17, 22, 26, 27
   and boron deficiency sensitivity 3:19
   and boron sensitivity 3:18, 20, 36
   and liming 1:94-96; 2:18
   and molybdenum 3:58
   and potassium 2:71
   sulfur deficiency in 2:70
   superphosphate 2:71
   zinc deficiency sensitivity 2:112
Allotropism 1:36
Aluminum 1:41
   neutralized by calcium (lime) 1:93, 96; 2:16, 19, 51
   toxicity of 1:51
Aluminum phosphate 1:41, 48
Ammonia 1:20, 23, 24, 34, 113; 2:35, 73
Ammonification 1:22, 93; 2:15
Ammonium 1:16, 17, 21, 23-25; 3:24, 35, 50
   ions 2:64
Ammonium molybdate 3:55, 58, 64
Ammonium nitrate 1:34; 2:73
Ammonium nitrogen 2:35
Ammonium sulfate 1:21, 34, 113; 2:35, 71, 74, 75, 78, 79, 126; 3:57
Anemia 2:83
Annuals 1:51, 65
Apatite minerals 1:41
Aphids 3:103
Apples
   and boron deficiency sensitivity 3:19, 31
Ascorbic acid 2:82
Asparagus
   and boron sensitivity 3:36
   and liming 1:94
   and sulfur 2:66
   zinc deficiency sensitivity 2:112
Asters
   and boron deficiency sensitivity 3:19
Auxims 2:99
Bacteria, rhizobia see Rhizobia bacteria
Banding fertilizer 1:26, 27, 43, 51, 54, 59, 78-81, 95, 108, 111; 2:29, 54, 75, 89, 106; 3:22, 58
Beans
   and boron 3:19
   boron deficiency in 3:15-17, 20, 37, 81, 82
   and boron excess 3:71
   and boron sensitivity 3:18, 20, 32, 35, 36
   and calcium 1:95
   calcium deficiency in 1:91, 92, 117; 3:77, 83
   and iron 2:119
   iron deficiency in 2:135
   and soil maggots 3:103
   magnesium deficiency in 2:13-16, 38, 100; 3:78-81
   manganese deficiency in 2:44-47, 61; 3:78, 92, 93
   molybdenum deficiency in 3:64, 77, 84
   phosphorus deficiency in 1:36-38, 61, 77, 78
   potassium deficiency in 1:63-71, 82, 83, 88; 3:79-84
   sulfur deficiency in 2:65
   and excess water 3:104
   yellows disease in 2:55
   zinc deficiency in 2:98-101, 115; 3:84, 92, 93
   zinc deficiency sensitivity 2:112
Beets
   and boron deficiency 3:19-23, 31, 37
   and boron sensitivity 3:36
   copper deficiency in 2:95
   iron deficiency in 2:135
   magnesium deficiency in 2:38; 3:80
   manganese deficiency in 2:61
   molybdenum deficiency in 3:64
   molybdenum sensitivity 3:51
   nitrogen deficiency in 1:34
   phosphorus deficiency in 1:61
   potassium deficiency in 1:88; 3:80
Biotin 2:68
Biotite 1:69
Black rot 3:109
Blight disease 3:116, 117
Blossom-end rot 3:118
Blowsand 3:132
Blue vitriol 2:87
Borax 3:15, 18, 20, 22, 25, 28, 37
   fertilizer 3:24
Bordeaux sprays 2:43, 80
Boric acid 3:24, 28, 35, 37
Boron 1:92, 97, 115; 2:14, 47, 49, 57; 3:15-48
   and calcium 3:16
   cautions regarding use 3:33, 34
   chemistry in plants 3:16-18
   chemistry in soil 3:19-21
   deficiency 3:17, 18, 20, 27, 35
   deficiency symptoms 3:16, 27, 30, 31
   defined 3:15
   excess 2:83; 3:17, 18, 21, 32
   fertilizer 2:84; 3:23, 24, 27, 29, 34-37
   fixation 3:27
   forms utilized by plants 3:22-25
   inorganic 3:20, 22
   methods of applying 3:28, 29
   organic 3:20, 22

recommended applications 1:139; 3:35
suggestions regarding 3:35, 36
summary of 3:37-48
supply, actions affecting 3:26, 27
toxicity 3:17, 18, 32
as weed killer 3:21, 33, 34
Boron salts 3:18, 21
Boron trioxide 3:24, 25, 29, 35
Brassicas
   molybdenum deficiency in 3:59, 60
   molybdenum sensitivity 3:51, 54
Broadcasting fertilizer 1:26, 51, 54, 80, 95, 108, 111; 2:29, 54, 74, 75, 89, 128; 3:22, 28, 58
Broccoli
   boron deficiency in 3:24, 27, 31, 37
   calcium deficiency in 1:117
   copper deficiency in 2:82, 83, 95
   iron deficiency in 2:135
   magnesium deficiency in 2:19, 22-25, 38
   manganese deficiency in 2:48, 49; 3:78
   and molybdenum 3:53
   molybdenum deficiency in 3:64
   nitrogen deficiency in 1:34
   phosphorus deficiency in 1:56, 61; 3:78
   potassium deficiency in 1:88
Brussels sprouts
   calcium deficiency in 1:117
   magnesium deficiency in 2:38
   molybdenum deficiency in 3:64
   potassium deficiency in 1:88
Bushbeans
   and molybdenum 3:58
Cabbage
   and bacterial disease in plant sap 3:115
   and black rot 3:109
   and boron 3:34
   and boron sensitivity 3:20, 36
   and boron deficiency 3:37
   and cabbage worms 3:103
   calcium deficiency in 1:117
   chlorine deficiency in 1:124
   copper deficiency in 2:80, 81, 83, 95
   and eelworms 3:102
   iron deficiency in 2:135
   and soil maggots 3:102
   magnesium deficiency in 2:19-21, 38; 3:75
   manganese deficiency in 2:50, 61; 3:94
   manganese excess in 2:57
   molybdenum deficiency in 3:49-52, 64
   and nitrogen 1:24
   nitrogen deficiency in 1:22, 35; 3:75
   phosphorus deficiency in 1:43, 44, 61; 3:75
   and potassium 1:72
   potassium deficiency in 1:71, 88
   and infertile soil 3:102
   sulfur deficiency in 2:67; 3:94
   sulfur in 2:65, 66
   and typhoon winds 3:108
   zinc deficiency in 2:107
Calcium 1:41, 90-119; 2:14-17, 21, 64
   and boron 3:16, 17
   cautions regarding use 1:113, 114
   chemistry in plants 1:91-94
   chemistry in soil 1:95-100
   and copper 2:92
   deficiency 1:92, 115
   deficiency symptoms 1:110, 111; 3:16
   defined 1:90
   excess symptoms 1:112
   fertilizers 1:117
   forms utilized by plants 1:101-104
   ions 1:98, 105; 2:20, 21, 26, 67
   and iron deficiency 2:127
   and magnesium 1:92, 96, 101-104; 2:85
   molybdenized 3:55
   and phosphorus 1:39
   and potassium 1:76
   recommended applications 1:139
   suggestions on use of 1:115, 116
   summary of 1:117-119
   supply, actions affecting 1:105-108
Calcium borate 3:20
Calcium carbonate 1:42, 90, 91, 97, 98, 101, 103, 112; 2:20-25, 27, 74
   boronated 3:24
Calcium chloride 1:102, 108, 112
Calcium cyanamide 1:20
Calcium hydroxide 1:101, 104, 106, 108; 2:80, 107
Calcium metaphosphate 1:46
Calcium nitrate 1:34, 108, 112
   boronated 3:24
Calcium oxide 1:94, 101, 104, 106; 2:25
Calcium oxolate 1:92
Calcium phosphate 1:41, 42, 49, 51
Calcium salt 1:42
Calcium sulfate 1:43, 108, 112; 2:65
Cantaloupe
   boron deficiency in 3:37
   calcium deficiency in 1:117
   magnesium deficiency in 2:33, 38
   molybdenum deficiency in 3:65
   potassium deficiency in 1:88
Carbohydrate metabolism
   and boron 3:16
   and sulfur 2:68
Carbohydrates 1:38, 92, 93; 2:15
Carbon 1:126-135
   cautions regarding use 1:134
   chemistry in soil 1:128
   cycle 1:128, 131, 135
   deficiency symptoms 1:132
   defined 1:126
   essential to plants, animals 1:127
   excess symptoms 1:133
   forms useful to plants 1:129
   hints regarding 1:135
   metabolism, and zinc 2:108
   supply, actions affecting 1:130
Carbon dioxide 1:103, 127-131, 134, 135; 2:25, 99
Carbon disulfite 2:66
Carbonic acid 2:99
Carbonic anhydrose 2:99
Carrots
   boron deficiency in 3:25, 26, 37

and boron sensitivity 3:36
calcium deficiency in 1:96, 97, 118
copper deficiency in 2:95
magnesium deficiency in 2:17-19, 38; 3:80
manganese deficiency in 2:51-53
molybdenum deficiency in 3:65
nitrogen deficiency in 1:21, 35
phosphorus deficiency in 1:61
potassium deficiency in 1:88; 3:80
seeds planted too thick 3:107
zinc deficiency sensitivity 2:112

Cation 1:69, 70, 137; 2:26, 85; 3:16
Cauliflower
and bacterial disease in plant sap 3:115
and boron 3:28-30
and boron deficiency 3:31, 37
and boron deficiency sensitivity 3:19
calcium deficiency in 1:118
iron deficiency in 2:135
and liming 1:94
magnesium deficiency in 2:26, 39
and molybdenum 3:54, 56
molybdenum deficiency in 3:51, 53, 55, 60, 65
nitrogen deficiency in 1:35
potassium deficiency in 1:88
and sulfur 2:66

Celery
and boron 3:33
and boron deficiency 3:31, 32, 38
boron deficiency sensitivity 3:19
calcium deficiency in 1:118
and liming 1:94
magnesium deficiency in 2:27, 39
manganese deficiency in 2:61
molybdenum deficiency in 3:51, 60, 65
nitrogen deficiency in 1:35
phosphorus deficiency in 1:61
potassium deficiency in 1:78, 88

Cell walls of plants
calcium and magnesium essential for 1:92; 2:15

Centigrade to Fahrenheit 1:141
Cereal crops
and potassium 1:67
zinc deficiency sensitivity 2:112

Chalk 1:90, 101, 102
Chard
molybdenum deficiency in 3:65

Chloride, stannous 2:52
Chlorine 1:41, 120-125, 137; 2:52
chemistry in plants 1:121
chemistry in soil 1:122
deficiencies, methods of diagnosing for 1:124
defined 1:120
recommended applications 1:125, 139
summary of 1:125
supply, sources of 1:123

Chlorine salts 1:122
Chlorophyll 1:14, 57, 135; 2:14, 31, 44, 57, 68, 82, 83, 99, 117, 118, 130, 131; 3:52, 59
Chloroplasts 2:82
Citric acid 2:124

Citrus fruit, trees
and copper 2:82
copper deficiency in 2:90, 91
iron deficiency in 2:116, 136
magnesium deficiency in 2:28
molybdenum deficiency in 3:57
nitrogen deficiency in 1:26
sulfur deficiency in 2:64, 76
zinc deficiency in 2:108, 109
and zinc sulfate 2:107

Clover
boron deficiency sensitivity 3:19
boron sensitivity 3:36
and liming 1:94
potassium deficiency in 1:76
zinc deficiency sensitivity 2:112

Coal 1:21; 2:73, 74
Coal ashes 1:106
Cobalt
and iron deficiency 2:127

Coke 2:66, 74
Colemanite 3:22, 23
Compost 1:59; 2:15; 3:111
Conversion see Fixation
Copper 1:97; 2:44, 56, 80-95
cautions regarding use 2:92, 93
chelated 2:87, 89, 94
chemistry in plants 2:81-84
chemistry in the soil 2:85, 86
deficiency 2:82, 86-89, 93
deficiency symptoms 2:83, 90, 91
defined 2:80
essential for plants 2:94
effect on soil 2:86
excess 2:83, 92
fertilizer 2:88, 90, 93, 94
hints on 2:94
and iron 2:117
and iron deficiency 2:127
methods of applying to plants 2:89
recommended applications 2:94
summary of 2:94, 95
supply, actions affecting 2:88

Copper chloride 2:87
Copper nitrate 2:87
Copper oxide 2:94
Copper sulfate 1:109; 2:86-89, 92, 94; 3:52
recommended applications 1:139; 2:84, 93

Corn
and boron 3:17, 26
boron deficiency in 3:38, 85, 86, 113
calcium deficiency in 1:100, 109, 110, 118; 3:86, 87
copper deficiency in 2:95
iron deficiency in 2:116, 135; 3:90
magnesium deficiency in 2:30, 31, 39; 3:79, 86
manganese deficiency in 2:54, 61, 105, 106; 3:89-91
and molybdenum 3:58
molybdenum deficiency 3:87
nitrogen deficiency in 1:14-17, 19, 35; 3:85, 113

and phosphorus 1:39, 43, 47
phosphorus deficiency in 1:40, 61
potassium deficiency in 1:63, 77-80, 82; 3:85, 113
and sulfur 2:66
and zinc 2:98
zinc deficiency in 2:99, 103-106, 110, 115; 3:85, 89-91
zinc deficiency sensitivity 2:112

Cotton
and boron sensitivity 3:36
and fertilizing practices 3:98
and nitrogen 1:16
potassium deficiency in 1:76
and low soil fertility 3:101
and sulfur 2:66
sulfur deficiency in 2:76

Creosote 3:111
Crop removal 1:23; 2:17, 73, 79
Crop rotation
and boron 3:32

Cucumbers
and boron 3:38
and boron deficiency 3:38, 39, 81
calcium deficiency in 1:97, 103, 118
magnesium deficiency in 2:28-32, 39
manganese deficiency in 2:61
and molybdenum 3:57
molybdenum deficiency in 3:65
nitrogen deficiency in 1:24, 35
nitrogen excess in 1:25
phosphorus deficiency in 1:61
potassium deficiency in 1:72-75, 89; 3:81

Curley top 3:116
Custom-made soil see Soil, custom-made
Cystine 2:67
Cytoplasm 1:39
Deficiencies, two or more 3:75-100
Diammonium phosphate 1:34, 60; 2:73
Diazinon 3:109
Dieback 2:91, 109
Dolomite lime see Lime
Drill-seeded crops
and molybdenum 3:58
Dry spot 2:44, 55
Eelworms 3:102, 112
Egg shells 1:117
Eggplant
and boron 3:36
Epsom salt 2:13, 36
Erosion 1:23, 76; 2:73, 77
Fahrenheit to Centigrade 1:141
Feldspar 1:136
Ferric iron 2:116, 120
Ferric sulfate 2:71, 79, 119, 124, 128, 129, 133, 134
Ferrous iron 2:116, 120
Ferrous sulfate 2:79, 119, 124, 128, 129, 133, 134
Fertility, low 3:101, 104-107, 112
Fertilizer
application of 3:113-115
excess 3:70-74

Mittleider pre-plant mix 3:134, 135

Field crops
and boron deficiency sensitivity 3:28
and zinc 2:107
zinc deficiency in 2:110
and zinc sulfate 2:105

Filberts
boron deficiency sensitivity 3:19

Fixation
boron 3:27
iron 2:123, 129
nitrogen 1:19; 3:50
phosphorus, phosphates 1:42-53, 80; 2:123; 3:54
potassium 1:68, 71, 72
zinc 2:105

Flowering plants, annual
boron sensitivity 3:36

Fluorapatite 1:41
Fluorine 1:41
Foliar sprays 1:27, 56, 75; 2:25, 30, 48, 49, 54, 75, 80, 89, 102, 106, 107, 112, 114, 119, 128, 129; 3:24, 27, 28, 35, 58

Forage crops
and copper sulfate 3:61
and molybdenum toxicity 3:61, 62

Fruit trees
and boron dusts, sprays 3:28
boron sensitivity 3:20, 36
iron deficiency in 2:126, 127, 130
and sodium borate 3:23
zinc deficiency in 2:109

Fruits
and boron 3:17

Gas, natural 2:73, 74

Grain crops
boron sensitivity 3:36
copper deficiency symptoms 2:90
nitrogen in 1:15
phosphorus in 1:41

Grapefruit
manganese deficiency in 2:53

Grapes
boron sensitivity 3:36
boron deficiency sensitivity 3:19
zinc deficiency sensitivity 2:112

Grasses
and boron 3:17
and low fertility 3:105
zinc deficiency sensitivity 2:112

Grasshoppers 3:103
Gray speck 2:44, 55
Greenhouse crops
and iron 2:133
Ground, rocky
made productive by water, fertilizer 3:101
Grow-boxes, garden 3:120-135
how to make 3:124-130
Gypsum 1:43, 102, 114, 115, 117; 2:74, 79; 3:133-135
Hay crops
phosphorus in 1:41
Hemoglobin 2:82, 83

138

Hops
  zinc deficiency sensitivity 2:112
Hormone movement
  and boron 3:16
Hydrogen ions 1:98, 2:20, 21, 26
Hydrogen sulfate 2:65
Hydrolysis 2:128
Hydroquinone 2:52
Hydroxides 2:124
Hydroxyol ions 2:48, 52, 53
Hydroxyapatite 1:41
Humus formation 1:93; 2:15
Illite 1:69
Iron 1:41, 48; 2:44, 45, 55, 57, 104, 116-136
  active 2:117
  chelates 2:119, 124-126, 128, 129
  chemistry in plants 2:117-119
  chemistry in soil 2:120-123
  and copper 2:82, 83
  deficiency 2:116-119, 121-124, 126, 127, 131, 133
  deficiency symptoms 2:119, 130, 131, 135, 136
  defined 2:116
  excess 2:119
  excess symptoms 2:132
  fertilizers 2:118, 122, 134
  fixation 2:123, 129
  forms utilized by plants 2:124, 125
  hints regarding 2:133
  ions 2:124
  methods of applying 2:129
  and molybdenum and nitrogen 2:119
  and molybdenum, nitrogen, and manganese 3:52
  neutralized by calcium (lime) 1:93, 96; 2:16, 19, 51
  recommended applications 1:139; 2:118, 133
  residual 2:117
  summary of 2:134-136
  supplements 2:119
  supply, actions affecting 2:126-128
  toxicity 2:132
  and zinc deficiency 2:99
  and zinc and manganese 2:128
Iron oxide 2:119, 121, 122, 125
Iron phosphate 1:41, 121
Iron sulfate
  recommended applications 1:139; 2:119
Kale
  and sulfur 2:66
Koalinite 1:69
Leaching 1:17, 18, 22, 23, 26, 66, 71, 73, 78, 85, 91, 95, 113; 2:17, 24, 45-47, 72-74, 77, 79, 86, 88; 3:21, 22, 33, 74
Leaf miners 3:104
Leaf spot 2:55
Legume crops
  and liming 1:95; 2:18
  and molybdenum 3:51, 54, 62
  and nitrogen 1:17, 20
  potassium deficiency in 1:82
Lemon
  iron deficiency in 2:117
Lentils
  and sulfur 2:65
Lettuce
  and boron 3:40
  boron deficiency in 3:38, 39
  boron sensitivity 3:36
  calcium deficiency in 1:101, 102, 110, 118
  chlorine deficiency in 1:124
  copper deficiency in 2:95
  and liming 1:94
  magnesium deficiency in 2:39
  manganese deficiency in 2:61
  molybdenum deficiency 3:65
  molybdenum sensitivity 3:51
  nitrogen in 1:15
  nitrogen deficiency in 1:35
  phosphorus deficiency in 1:61
  potassium deficiency in 1:89
  sulfur deficiency in 2:68
Lily family
  plants in, require lots of sulfur 2:66
Lime (limestone) 1:42, 90, 92, 95, 102, 103, 111, 115; 2:22, 25, 27, 35, 50, 53, 105, 118, 120, 121, 130; 3:26, 113, 133-135
  agricultural 1:100, 108, 115, 117; 2:24, 27, 74
  applying to land 1:107, 2:29
  burned 1:101, 103, 108, 117
  calcitic 1:90, 2:28
  caustic 1:103, 104
  dolomite 1:90, 101, 102, 108, 115, 117; 2:17-25, 28, 29, 34, 37
  excess 1:100
  hydrated 1:101, 104, 109
  lump 1:103
  methods of applying 1:108, 109; 2:29, 30
  and molybdenum availability 3:53
  quicklime 1:103, 117
  recommended applications 1:99, 100
  slacked 1:104, 108, 117; 2:80
  unslacked 1:103
Liming 1:51, 53, 72, 91, 94, 96-99, 102, 104-106; 2:19-22, 26, 47, 51, 58, 72; 3:28, 54, 56, 57
Little leaf 2:108, 109
Lodging 1:78
Maggots, soil 3:102, 103, 109
Magnesium 1:92, 96; 2:13-42
  calcium, potassium, and boron, balance in plants between 2:14
  cautions regarding use 2:34, 35
  chemistry in plants 2:14-16
  chemistry in soil 2:17-23
  and copper 2:85, 92
  deficiency 2:35, 36
  deficiency symptoms 2:31, 32
  defined 2:13
  excess 1:85; 2:33
  fertilizers 2:37
  forms of useful to plants 2:24, 25
  functions of 2:15
  hints regarding 2:36
  ions 1:98, 105; 2:20, 21, 26
  summary of 2:37-42

supply, actions affecting 2:26-28
and zinc 2:99
Magnesium borate 3:20
Magnesium carbonate 1:98, 103, 106; 2:13, 20-25
Magnesium oxide 1:106; 2:13, 16, 18, 21, 24-30, 34, 37
Magnesium sulfate 2:13, 18, 21, 23-26, 29, 30, 34, 37, 65, 71, 78, 79; 3:134
    recommended applications 1:139; 2:36
Manganese 2:43-63, 104
    cautions regarding use 2:58
    chelates 2:60
    chemistry in plants 2:44
    chemistry in soil 2:45-47
    and copper 2:83
    cycle 2:52
    deficiency 2:50, 51, 58
    deficiency symptoms 2:55, 56
    defined 2:43
    excess symptoms 2:57
    fertilizers 2:60
    forms utilized by plants 2:48, 49
    ions 2:47
    and iron 2:117
    and iron deficiency 2:127
    and iron, nitrogen, and molybednum 3:52
    and iron and zinc 2:128
    methods of applying to plants 2:54
    and molybdenum availability 3:53
    and molybdenum deficiency 3:57
    neutralized by calcium (lime) 1:93-97; 2:16, 19
    recommended applications 2:49
    suggestions regarding 2:59
    summary of 2:60-63
    supply, actions affecting 2:50-53
    toxicity of 1:51; 2:52, 58, 59
    and zinc 2:99
Manganese oxide 2:43-50, 60
Manganese salts 2:53
Manganese sulfate 2:43, 48, 49, 54, 60
    recommended applications 1:139; 2:53, 59
Manganic-manganese 2:46, 58
Manganous-manganese 2:46, 51, 58
Manure crops, green 2:113
Manures 1:21, 34, 59, 70, 74
Marl 1:101-103
Marsh spot 2:55
Meals 1:21, 34, 59, 70
Measure, units of 1:141
Melons
    boron deficiency in 3:38, 40, 41, 88
    calcium deficiency in 1:98, 100
    magnesium deficiency in 2:32
    and molybdenum 3:58
    molybdenum deficiency in 3:58-61, 88
    and potassium 1:80
    potassium deficiency in 1:73, 89
    and wilting 3:105
Methionine 2:67
Mica 1:69
Miscellaneous symptoms 3:101-119

Milo
    and Pyrethrum 3:110
Mittleider method 3:123
Mittleider pre-plant fertilizer mix 3:134, 135
Molybdate 3:53
Molybdenosis 3:50, 52, 61
Molybdenum 1:97; 2:19; 3:49-68
    cautions regarding use 3:62
    chemistry in plants 3:50-52
    chemistry in soil 3:53, 54
    deficiency of 1:51; 2:51; 3:50-56, 59
    deficiency symptoms 3:60
    defined 3:49
    excess symptoms 3:61
    in family gardens 3:63
    fertilizers 3:55, 63, 64
    forms utilized by plants 3:55
    in greenhouses 3:63
    ions 3:56
    and iron and nitrogen 2:119
    and iron, nitrogen, and manganese 3:52
    and manganese 3:57
    methods of applying to plants 3:58, 59
    and nitrogen 3:60, 62
    and phosphorus 3:57, 62
    powder 3:51
    recommended applications 1:139; 3:62, 63
    suggestions regarding 3:63
    and sulfur 3:62
    summary of 3:64-68
    supply, actions affecting 3:56, 57
    toxicity 3:50, 52, 53, 61
Molybdenum salts 3:50, 52
Molybdic acid 3:52, 55, 58, 64
Monoammonium phosphate 1:60
Montmorillinite 1:69, 71
Mottle leaf 2:109
Muck see Peat
Mulches 1:74; 2:53; 3:110
Mustard family
    plants in, require lots of sulfur 2:66
    zinc deficiency sensitivity 2:112
Neutralizing value 1:102, 105, 106; 2:25, 27
Nickel
    and iron 2:117
Nitrate 1:16-19, 21-25, 32; 2:35
Nitric acid 1:19; 2:65
Nitrogen 1:13-35, 97; 2:16-19, 35, 69, 74, 78
    and borax 3:28
    and boron 3:17
    cautions regarding use 1:32
    chemistry in plants 1:14, 15
    chemistry in soil 1:16-18
    and copper 2:83
    and copper deficiency 2:88, 92
    cycle 1:16
    decaying processes require 1:16
    deficiency 2:76, 114; 3:51, 52
    deficiency symptoms 1:29, 30
    defined 1:13
    dependence of life processes on 1:14
    excess symptoms 1:31
    fertilizers 1:34; 2:68, 113

fixation 1:19; 2:68; 3:50
fixed 1:20
forms plants can utilize 1:19-21
gas 1:16-19, 93; 2:15
hints regarding 1:33
and iron deficiency 2:127
and iron and molybdenum 2:119
and iron, molybdenum, and manganese 3:52
metabolism, and boron 3:16
methods of applying to soil 1:26-28
and molybdenum 3:60, 62
organic 20
oxides 1:19, 20
recommended applications 1:23, 139
regulator in plants 1:15
soil loss of 1:14
summary of 1:34, 35
supply, actions affecting 1:22-25
Nitrous acid 1:19
NPK fertilizer 2:71
Nut trees
  boron sensitivity 3:20
Oil 2:74
Okra
  manganese deficiency in 2:55
  nitrogen deficiency in 1:27
Onions
  boron deficiency in 3:38
  boron sensitivity 3:20
  copper deficiency in 2:95
  magnesium deficiency in 2:39
  manganese deficiency in 2:43
  molybdenum deficiency in 3:62
  nitrogen deficiency in 1:35
  phosphorus deficiency in 1:61
  potassium deficiency in 1:89
  and sulfur 2:66
  zinc deficiency in 2:109
  zinc deficiency sensitivity 2:112
Oranges
  sulfur deficiency in 2:69
Orchard crops
  and zinc 2:100, 106
  zinc deficiency in 2:104
Ornamental crops
  boron deficiency sensitivity 3:19
  iron deficiency in 2:126, 133
Oxolic acid 1:92
Oyster shells 1:101-103, 117
Pahala blight 2:55
Palisade cells 2:108
Pansies
  and excess fertilizer 3:74
  nitrogen deficiency in 1:27, 28, 49
  phosphorus deficiency in 1:49
  and excess salt 3:73
Papaya
  and lime 1:105
Pasture crops
  and molybdenum salts 3:50
Peaches
  and boron 3:26
Peanuts
  potassium deficiency in 1:81
Pears
  and boron deficiency 3:31
  boron deficiency sensitivity 3:19
Peas
  boron deficiency in 3:38, 42
  calcium deficiency in 1:118
  copper deficiency in 2:85-88, 95
  magnesium deficiency in 2:39
  and molybdenum 3:58
  molybdenum deficiency in 3:65
  phosphorus deficiency in 1:61
  potassium deficiency in 1:82, 83, 89
  and sulfur 2:65
  zinc deficiency in 2:115
  zinc deficiency sensitivity 2:112
Peat 1:65, 70, 76, 103; 2:92
Peat moss 3:59, 132
Peat scows 3:61
Pecan trees
  zinc deficiency in 2:109
Pectate 1:91
Pectic substances metabolism
  and boron 3:16
Pectin 1:91
Peppers
  and boron sensitivity 3:36
  magnesium deficiency in 2:13, 33, 34, 42; 3:78
  and phosphorus 1:45, 47
  phosphorus deficiency in 1:46, 61; 3:78
  sulfur deficiency in 2:66
Perlite 3:59, 132
Petioles 3:30
pH of soil 3:69-74
  re boron 3:24-28
    calcium 1:95-100, 105, 114
    copper 2:83, 85, 92
    iron 2:120, 121, 125, 127-129
    magnesium 2:19, 20, 26
    manganese 2:45, 47, 50, 51, 58, 59
    molybdenum 3:53, 56, 57
    nitrogen 1:22, 23
    phosphorus 1:43, 48-53
    sulfur 2:69, 74, 75, 78
    zinc 2:100, 101, 104, 105, 113
Phosphate fertilizer 1:42, 47, 51, 53, 57; 2:121
Phosphates 2:124; 3:53
  rock 1:46, 104; 2:75
Phosphoric acid 1:46, 55, 56, 60
Phosphorus 1:36-62, 97, 99; 2:19, 122, 123
  accumulations in soil 1:50
  and boron 3:17
  and calcium 1:39
  chemistry in plants 1:38-40
  chemistry in soil 1:41-43
  common yellow 1:37
  and copper 2:92
  and copper deficiency 2:93
  deficiency symptoms 1:57, 58
  defined 1:36, 37
  fertilizers 1:60; 2:113
  fixation 1:42-53, 80, 123; 3:54

141

forms plants can utilize 1:44-46
growth requirements 2:99
hints, suggestions 1:59
and iron 2:118
and iron deficiency 2:127
methods used to apply to plants 1:54-56
and molybdenum 3:57, 62
natural supplies 1:41
and nitrogen and potassium 1:50
recommended applications 1:139
red-powdered 1:37
and sulfur 2:64
summary of 1:60-62
supply, actions affecting 1:47-53
supply per acre (inorganic) 1:37
and zinc 2:105
and zinc, iron, and manganese 1:51, 104, 128
Phosphorus pentoxide 1:37, 40, 48
Photosynthesis 1:30, 38; 2:81
Potash 1:63, 64
    caustic 1:63
    muriate of 1:63, 71, 88
    nitrate of 1:63, 75, 88
    sulfate of 1:63, 88
Potash salts 3:32
Potassium 1:63-89, 92; 2:14, 16, 85, 122, 123
    and borax 3:28
    and boron 3:17
    and calcium 1:76
    cautions regarding use 1:85, 86
    chemistry in plants 1:65-68
    chemistry in soil 1:69-73
    and copper 2:92
    deficiency 2:114
    deficiency symptoms 1:67, 68, 82, 83
    defined 1:63, 64
    excess 1:85
    excess and manganese deficiency 1:84
    excess symptoms 1:84
    fertilizer 1:75-78, 88
    fixation 1:68, 71, 72
    forms of in soil 1:80
    forms utilized by plants 1:74, 75
    ions 1:70, 71, 105; 2:26, 64
    and iron deficiency 2:127
    and lime 1:94
    methods of applying 1:80, 81
    and nitrogen and phosphorus 1:77
    recommended amounts 1:68, 139
    and soil moisture 1:86
    suggestions regarding 1:87
    summary of 1:88, 89
    supply, actions affecting 1:76-79
    supply per acre (farmland) 1:70
    supply per acre (mineral soil) 1:73
    toxicity 1:84
Potassium carbonate 1:63, 64
Potassium chloride 1:73, 87, 122; 2:75; 3:134
Potassium nitrate 1:34, 87; 2:73
Potassium oxide 1:64
Potassium sulfate 1:69, 87; 2:71, 75, 78; 3:134
Potato scab 1:91
Potatoes

    boron deficiency in 3:38
    and boron excess 3:32
    boron sensitivity 3:20, 36
    calcium deficiency in 1:118
    and compost only 3:111
    copper deficiency in 2:89, 95
    iron deficiency in 2:135
    magnesium deficiency in 1:51; 2:35, 36, 40
    manganese deficiency in 1:51; 2:56, 57, 61
    manganese excess in 2:57
    molybdenum deficiency in 3:66
    molybdenum sensitivity 3:51
    nitrogen deficiency in 1:29, 35
    phosphorus deficiency in 1:49-51, 62
    and potash fertilizers 3:32
    potassium deficiency in 1:78, 89
    zinc deficiency in 2:115
    zinc deficiency sensitivity 2:112
Proteins 2:81
Protoplasm 1:30, 38
Pyrethrum 3:110
Radishes
    boron deficiency in 3:38, 43
    boron sensitivity 3:36
    calcium deficiency in 1:118
    and fertilizer 3:115
    magnesium deficiency in 2:40
    molybdenum deficiency in 3:63, 66
    nitrogen in 1:15
    phosphorus deficiency in 1:62
    potassium deficiency in 1:84, 89
    and sulfur 2:66
Rape
    molybdenum sensitivity 3:51
    and sulfur 2:66
Rhizobia bacteria 1:20, 24, 93; 2:15, 18; 3:50, 54, 58, 62
Rocky ground see Ground, rocky
Rosette 2:109; 3:30, 38, 41, 44
Ruminants
    and molybdenum toxicity 3:52, 61, 62
Rutabaga
    and boron deficiency 3:31
Safflower
    zinc deficiency sensitivity 2:112
Salinity 2:35; 3:69-74
Salt absorption
    and boron 3:16
Salt burn 3:109
Salt excess 3:71-73
Scab disease 3:116
Seawater
    and boron 3:15
Side-dressing fertilizer 1:25, 27, 55, 79, 81; 2:30, 54, 75, 103, 107; 3:28
Silicates, hydrous 1:69
Slag 1:101, 102, 104; 2:103
Sodium 1:98, 136-138; 2:21
    defined 1:136, 137
    essential to animals 1:137
    excess symptoms 1:138
    ions 1:105; 2:26
    nonessential to plants 1:137

Sodium borate 3:20, 23, 25, 28, 34, 134
    and minimum boron crop requirement 3:18
    recommended applications 3:35
Sodium chloride 1:136
Sodium molybdate 3:55, 58, 64
Soil
    acid 1:41, 42, 48, 52, 55, 65, 69, 73, 76,
        94-99, 103, 115; 2:17, 18, 26, 45, 47, 50,
        53, 59, 78, 79, 84-88, 92, 99, 105, 122, 126,
        127; 3:26, 27, 35, 56, 57
    alkaline 1:24, 41, 42, 48, 49, 52, 54, 95-98,
        115, 137; 2:17, 18, 21, 34, 35, 45, 47, 51,
        53, 58, 59, 69, 78, 84, 92, 99, 104, 118,
        122, 126, 127; 3:23, 26, 27, 56, 101
    arid 2:126; 3:18, 19
    calcareous 1:41, 42, 70, 76, 92, 120
    clay 1:97, 105; 2:19, 34, 86, 121; 3:24, 26
    colloids 3:56
    custom-made 3:120-122, 131-135
    igneous rock 3:21
    infertile 3:102
    marine sediment 3:21
    mineral 1:69, 126; 3:54
    neutral 2:18, 34, 53, 78, 126; 3:35, 56
    noncalcareous 2:104
    organic 1:65, 76; 2:50, 51, 93; 3:24
    pH of see pH of soil
    saline 1:98, 137; 2:21
    sandy 1:65, 76, 97, 105; 2:19, 34, 86, 92, 121;
        3:22, 26
    silt 1:97; 2:19, 121; 3:26
    weathered 2:86
Solubar 3:24, 25, 37
Sorghum
    zinc deficiency sensitivity 2:112
Soybeans
    calcium deficiency in 1:106-109
    and fertilizer deficiency 3:76, 97
    and molybdenum 3:58
    phosphorus deficiency in 3:77
    potassium deficiency in 1:85, 86; 3:77
    and weed control 3:97
    and weeds 3:112
    zinc deficiency in 2:109
    zinc deficiency sensitivity 2:112
Spinach
    manganese deficiency in 2:58
    molybdenum deficiency in 3:66
    molybdenum sensitivity 3:51
    yellows disease in 2:55
Squash
    and boron 3:44
    calcium deficiency in 1:104
    copper deficiency in 2:81
    magnesium deficiency in 2:39, 40
    manganese deficiency in 2:58-61; 3:89, 93
    molybdenum deficiency in 3:63, 66
    potassium deficiency in 1:85
    zinc deficiency in 3:89, 93
Stannous chloride 2:52
Strawberries
    and boron sensitivity 3:18, 32, 35, 36
    and excess boron fertilizer 3:71

    and iron 2:118
Subsoil
    and copper 2:85
    and zinc 2:100, 101
Sudan grass
    zinc deficiency sensitivity 2:112
Sugar beets
    and boron 3:17, 26
    boron deficiency sensitivity 3:19
    boron sensitivity 3:20
    and phosphorus 1:36
Sugar cane 2:55
Sugar translocation in plants
    and boron 3:16
Sulfate 2:65, 67, 69; 3:53
    of ammonia 3:134
    fertilizer, and molybdenum deficiency 3:56
    ions 2:65, 75
    of potash 2:79
Sulfhydryl 2:67
Sulfur 1:43, 113; 2:44, 47, 48, 55, 64-79, 126
    cautions regarding use 2:77, 78
    chemistry in plants 2:66-69
    chemistry in the soil 2:70, 71
    deficiency symptoms 2:68, 76, 77
    defined 2:64-66
    essential to all life 2:64
    excess symptoms 2:77
    fertilizer 2:79
    forms utilized by plants 2:71, 72
    ions 2:69, 71
    methods of applying to plants 2:74, 75
    and molybdenum 3:62
    oxidation 1:93; 2:15
    recommended applications 1:139; 2:69, 70,
        77
    suggestions regarding 2:78, 79
    summary of 2:79
    supply, actions affecting 2:72-74
Sulfur dioxide 2:64, 67, 69, 72, 79
Sulfuric acid 1:42; 2:65, 75; 3:53
Sunflowers
    boron deficiency sensitivity 3:19
    potassium deficiency in 1:86
Superphosphates 1:41-46, 50, 59, 60; 2:71, 75,
        78, 79; 3:134
    and boron 3:28
    boronated 3:24
    molybdenized 3:58
Sweet potatoes
    and boron 3:44
    boron deficiency in 3:39, 95
    calcium deficiency in 3:95
    and low fertility 3:106
    and fertilizer 3:99
    fertilizer deficiency in 3:96
    magnesium deficiency in 2:37, 38, 40; 3:95
    nitrogen deficiency in 3:76, 95
    and nutrient deficiency 3:96
    phosphorus deficiency in 1:62; 3:76, 95
    and plant food deficiency 3:96
    potassium deficiency in 3:95
    and weeds 3:106

Swiss chard
    and boron 3:34, 35
Symptoms, miscellaneous 3:101-119
Tapioca
    phosphorus deficiency in 1:52
Taro
    potassium deficiency in 1:87
    and scab disease 3:116
Teart disease 3:61
Thiamine 2:68
Tobacco
    sulfur deficiency in 2:76
Tobacco mosaic 3:118, 119
Tomatoes
    and blight disease 3:117
    and boron 3:18, 46, 47
    boron deficiency in 3:39, 45
    boron sensitivity 3:18, 20, 36
    calcium deficiency in 1:110, 119
    chlorine deficiency in 1:124
    copper deficiency in 2:90, 95
    and curley top 3:116
    and excess fertilizer 3:74
    and fertilizer deficiency 3:104
    and iron 2:121, 122, 136
    iron deficiency in 2:120
    magnesium deficiency in 2:40, 41
    manganese deficiency in 2:61-63
    and molybdenum 3:67
    molybdenum deficiency in 3:64-66
    molybdenum sensitivity 3:51
    nitrogen deficiency in 1:30-33, 35
    phosphorus deficiency in 1:53-55, 62
    potassium deficiency in 1:78, 87-89
    sulfur deficiency in 2:70, 71
    and sunburn 3:109
    and tobacco mosaic 3:118, 119
    and lack of variety 3:108
    and wilt disease 3:115, 117
    zinc deficiency in 2:97, 110, 111, 115
    zinc deficiency sensitivity 2:112
Top-dressing fertilizer 1:25, 68, 78-81; 2:30, 45, 54, 103, 107; 3:28, 69
Topsoil
    and copper 2:85
Tourmaline 3:19
Tricalcium phosphate 1:46
Tuber crops
    and potassium 1:67
Turnips
    boron sensitivity 3:36
    and sulfur 2:66
    sulfur deficiency in 2:76
2-4-D 3:110
Urea 1:34; 2:73
Vegetable crops
    and boron 3:17, 27
    and boron sprays, dusts 3:28
    and molybdenum 3:51
    nitrogen deficiency in 1:22
    and phosphorus 1:42
    and potash fertilizers 3:32
    and sodium borate 3:23
    zinc deficiency in 2:96, 97, 107
    and zinc sulfate 2:105, 107
Vermiculite 1:69, 71
Volitalization
    and nitrogen 1:18
Walnuts
    boron deficiency sensitivity 3:19
Water metabolism
    and boron 3:16
Watercress
    chlorine deficiency in 1:124
Watermelons
    boron deficiency in 3:42
    manganese deficiency in 2:54
Weeds 3:105, 106, 112
Wheat
    and blight disease 3:116
    nitrogen in 1:33
    nitrogen deficiency in 1:34
Whiptail disease 3:49, 51, 59, 60
White streak 2:55
White tip disease 2:90, 91
Wilt disease 3:117
Worms, cabbage 3:103
Yellow disease 2:44, 55, 109
Yellow spot disease 3:49
Yellow tip disease 2:90, 91
Zinc 1:97; 2:44, 51, 55, 56, 96-115
    cautions regarding use 2:112, 113
    chelated 2:102, 103, 106, 114
    chemistry in plants 2:97-99
    chemistry in soil 2:100, 101
    deficiency 2:101-107, 114
        management practices re 2:113
    deficiency symptoms 2:108-110
    defined 2:96
    essential for plants, animals 2:96
    excess symptoms 2:111
    fertilizers 2:103, 114
    fixation 2:100, 101, 105
    forms utilized by plants 2:102, 103
    and iron 2:117
    and iron deficiency 2:127
    and iron and manganese 2:128
    methods of applying to plants 2:106, 107
    recommended applications 2:99
    suggestions regarding 2:114
    summary of 2:114, 115
    supply, actions affecting 2:104, 105
    toxicity 2:105, 113
Zinc carbonate 2:103
Zinc dusts 2:107
Zinc oxide 2:103, 114
Zinc phosphate 2:103
Zinc salts 2:105
Zinc sulfate 1:139; 2:102, 103, 106, 107, 114
Zucchini
    and excess fertilizer 3:70
    and typhoon winds 3:107
    and virus disease 3:116, 119